Alluvial Fans & Fan Deltas

A Guide to Exploration for Oil and Gas

Alluvial Fans & Fan Deltas

A Guide to Exploration for Oil and Gas

Gordon S. Fraser
Indiana Geological Survey

Lee Suttner
Indiana University

International Human Resources Development Corporation
Boston / Houston / London

Library of Congress Cataloging-in-Publication Data

Fraser, Gordon S.
Alluvial Fans & Fan Deltas: a guide to exploration for oil and gas.

Bibliography: p. 199
1. Petroleum—Geology. 2. Gas, Natural—Geology.
3. Alluvial Fans. I. Suttner, Lee J. II. Title.
TN870.5.F73 1986 553.2'8 86-10514
ISBN 0-88746-096-8

Contents

Preface

In selecting and organizing the subject matter for this book we have tried to look at alluvial fans and fan deltas through the eyes of an explorationist. We treat topics in the order in which exploration in a frontier-type setting might proceed. First, we focus on the basic conditions which must be met for alluvial fans to form. This permits identification of the tectonic settings where one might search for ancient alluvial-fan deposits and allows cataloguing of the properties of various tectonic settings which control fan morphology and facies.

In Chapters 2 and 3, we look at the morphology, processes and environments of alluvial fans. Processes, and the areal arrangement of environments, define facies and facies stacking. The concepts which are presented will be useful in the demonstration of the actual presence of alluvial fan facies within a sedimentary basin whose tectonic setting would suggest the probable existence of an ancient fan.

In Chapter 4, we assume that an alluvial fan system has been identified in the subsurface. We develop an idealized model that details the fan architecture that would form in response to a single tectonic episode. We show how this model can be used by an explorationist to establish relative position within a fan complex from limited subsurface data and to predict the physical character of the sediments in areas of the complex where little or no data are available.

Chapter 5 focuses on variants of the idealized and general model developed in Chapter 4. We examine causes of variance and show how the causes can work separately and in combination to produce derivative fan models. Our goal is to provide the basis for generation of detailed predictive models for any alluvial fan complex, regardless of its tectonic setting, parent rock, climate, size, and evolutionary history.

The ideas developed in Chapters 4 and 5 are based, in part, on deductive reasoning. A large volume of literature is available regarding the responses of source area drainage basins to changes in extrinsic factors, particularly climate. Less is known about how changes in source area response affect fan evolution, and virtually nothing is available that describes their effects on marginal areas of fan deltas. The literature that is available offers some intriguing hints about the complex responses that may be triggered on fans by changes in extrinsic stress, but until more detailed data are available, Chapters IV and V should be considered as conceptual frameworks rather than rigid dogma.

In the last chapter we present a discussion of the variables controlling diagenesis of alluvial fan sediments. We highlight those aspects of diagenesis which pose the greatest potential significance in controlling reservoir properties, most notably porosity and permeability.

Because of the multitude of factors influencing them, the morphology and resultant facies distributions of alluvial fans are exceedingly complex. Fans, therefore, are not prone to analysis through counterpart models, largely because they cannot be characterized by simple hierarchical arrangements of facies. Because of this we emphasize processes and variables which affect the alluvial-fan system as a whole. We attempt to convey throughout the book our perception of the complexity of alluvial-fan architecture. Only in this manner will readers gain the flexibility in their thinking about alluvial fans to adapt whatever general models do exist to their own unique situation.

This book is an outgrowth of a series of lectures on alluvial fans and fan deltas presented by the authors to the Sohio Petroleum Company in May, 1985. Sohio graciously has permitted us to publish this book which is a revised and expanded version of the text written to accompany the lecture series. We thank them for this permission.

We gratefully acknowledge the generous help of several individuals in the preparation of this book. Foremost, is Kay Kinder who typed the entire manuscript. Jim Tolen and Barbara Hill were responsible for the bulk of the drafting and photography, respectively. Marcus Johnson assisted in literature research that resulted in the comprehensive bibliography on alluvial fans which supplements the text and provided meticulous help on several other aspects of the editing and production of the manuscript.

Lee J. Suttner
Gordon S. Fraser
Bloomington, Indiana

1

An Introduction to Alluvial Fans

DEFINITIONS

Alluvial fans have been defined on the basis of (i) their morphology, (ii) their composition, (iii) the conditions and processes which form them, and (iv) their setting. In order to appreciate fully the potential complexity of alluvial fans and their deposits, a comprehensive definition based on all of the above criteria is preferred. Therefore, we consider *alluvial fans* to be conical, lobate or arcuate accumulations of mostly coarse-grained sediment, deposited by both water and gravity-induced density flow, and which extend from a mountain front or an escarpment across an adjacent lowland. *Alluvial cones* are small (<1 km^2) and often isolated alluvial fans, likely to show very small preservation probability. *Scree* or *talus piles* are non-conical accumulations of coarse detritus deposited by gravity sliding. They are characterized by an absence of water-laid deposits, an important characteristic in their distinction from fan deposits in the ancient.

We emphasize a distinction between alluvial fans and alluvial aprons or alluvial plains which are extensive regional accumulations (1000's km^2) of mostly alluvial and eolian sediments associated with more distant areas of high relief. An example of an alluvial plain sequence would be the Cenozoic High Plains deposits of the western interior. Alluvial plains develop a different facies architecture than alluvial fans, and, as a result, we expect that alluvial-fan facies models often are applied incorrectly to ancient alluvial-plain deposits. A more complete discussion of alluvial plain facies models is found in Rust (1978). A remarkably detailed facies analysis of an ancient alluvial plain system making up the Ogallala Formation is presented by Seni (1980).

Knowledge of down-dip environments which fringe alluvial fans is important to understanding distal facies relationships and in evaluating potential hydrocarbon source rocks. In most extensional basins, alluvial fans coalesce to form bajadas, piedmont alluvial plains or compound alluvial fans which may show complex spatial association with basin axial environments including swamps, lakes, eolian plains and the major trunk drainage system in the basin. In coastal settings, alluvial fans can prograde into the marine environments to form fan-deltas. Fan deltas are characterized by close and complex juxtaposition of fan, coastal and offshore marine facies.

TECTONIC SETTING

Alluvial fans are found in virtually all climates, form in areas characterized by high, yet a relatively wide range in relief, and contain detritus derived from a broad range of parent rocks. Essential to their development is an abundant sediment yield and a setting where streams transporting the detritus emerge from confined valleys on to open plains and undergo a sharp reduction in gradient. In such cases, a stream reacts by depositing sediment abruptly because of the dual effects of its loss in competence and desire to re-establish equilibrium by re-steepening its gradient. Abrupt deposition diverts the stream; repetitive diversion, constricted by the locally linear mountain front, results in the fan morphology.

Because of the essential need for large sediment yield, alluvial fans are associated traditionally with arid or semi-arid climates where relative absence of vegetation promotes slope instability. However, variables other than climate control sediment availability, including parent rock and slope gradient. Consequently, so-called humid or wet alluvial fans are expected and are known to occur in humid climates where sediment yield is great. Humid fans also show a different facies architecture than dry fans; and, facies models for one are not applicable to the other as we shall emphasize later.

The importance of the need for a sharp reduction in stream gradient at the apex of the fan has not been fully appreciated in predicting tectonic settings for basins where thick, alluvial fan deposits should or should not be found. Although we do not attempt quantitatively to define a minimum abrupt gradient reduction needed for alluvial fan generation, we feel that most basins bounded by low-angle faults or no faults will be characterized by alluvial plain or apron deposition. Conversely, basins bounded by high angle faults, reverse, normal, or transcurrent, are likely settings for thick alluvial-fan accumulations.

Given the link between the nature of the basin-boundary structure and presence or absence of alluvial fans, it is possible to predict the tectonic settings of basins in which alluvial fans might or might not be found. Knowing tectonic setting, the first-order extrinsic controls on the alluvial fan composition and facies structure also can be described. These first-order controls include the precise nature of the bounding fault, which in turn affects facies stacking in alluvial fans (Fig. 1.1), major environments associated with the alluvial fan, and the thermal/burial history of the basin which affects diagenesis. We have attempted to catalogue this information in Table 1.

The organization of Table 1 is based on the widely accepted view that sedimentary basins can be grouped on the basis of crustal motion which affects their origin, namely convergence, divergence or strike-slip. There is little difficulty in relating crustal divergence to local crustal extension and thus high-angle normal faulting and alluvial-fan formation. Normal faulting in such settings is repetitive and is of the younger on older or step-back variety. The latter is an important factor in controlling facies megasequences in alluvial fans as we shall demonstrate later.

Crustal divergence may be mature or immature in development depending on the degree of propagation of rifting. Basins in areas of mature extension are complex because they can be partitioned by horsts or topographic barriers formed by rotated blocks (Fig. 1.2). Alluvial-fan fill will be temporally continuous, but spatially discontinuous. Basins in areas of immature divergence are characterized by alluvial fan deposition during incipient stages of their development in intraplate settings (Fig. 1.3). Along plate margins, immature divergence forms aulacogens. Alluvial fans can form early in the aulacogen's history and also possibly later if the basement structures are reactivated by crustal compression (Fig. 1.4). Compression-caused reactivation of basement structures, which were initially formed during crustal extension, may be the origin of some of the major foreland block uplifts of Pennsylvanian age in Colorado (e.g. Ancestral Rockies), Texas and Oklahoma. In general, in each of the divergent basin settings, alluvial fan facies will not have lateral marine facies equivalents, but structural juxtaposition of marine and non-marine facies is possible. Diagenetic

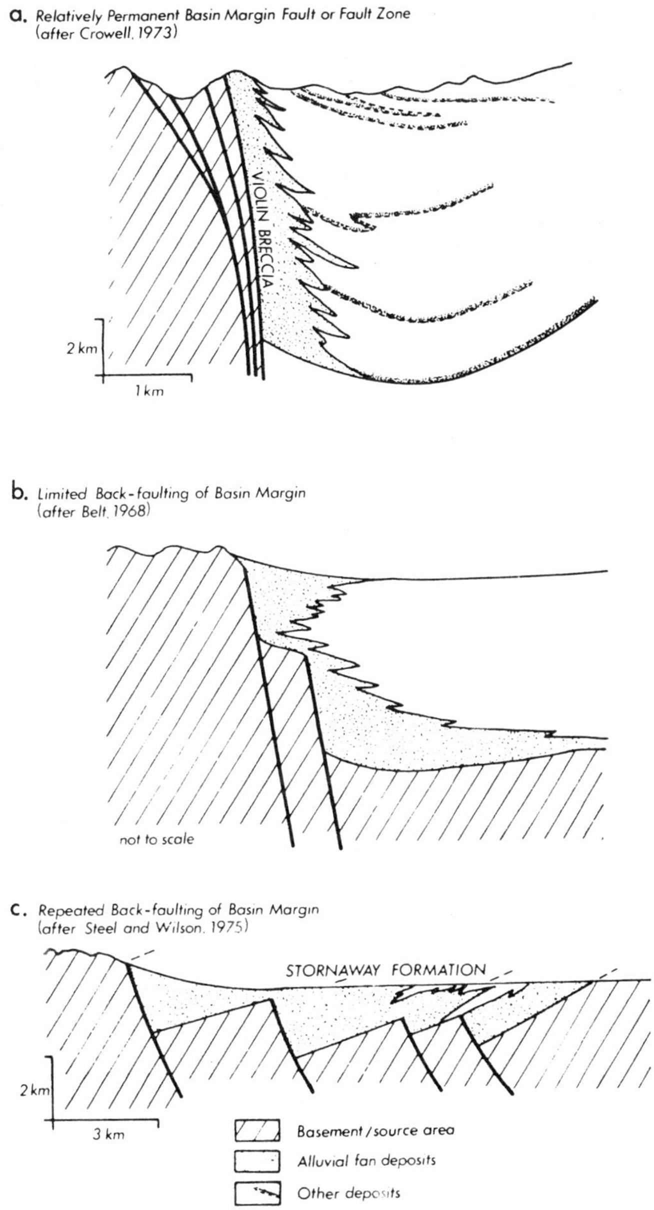

Figure 1.1 Contrasting facies stacking sequences and geometries of alluvial fan deposits associated with differences in basin-margin fault control (after Heward, 1978b).

	BASIN TYPE		BASIN BOUNDING FAULT	FACIES/ENVIRONMENTS ASSOCIATED WITH THE ALLUVIAL FANS	THERMAL/BURIAL HISTORY
CONVERGENT	Intracontinental arc	Volcano-tectonic basins	Normal, limited step-back	Alluvial, paludal, lacustrine	Highly variable--generalization difficult. Heat-flow ranges from high in forearc and intracontinental arc to low in back-arc setting. Immature detritus promotes cementation by clay, quartz and feldspar.
		Metamorphic core complex basins	Normal, step-back		
	Back-arc extension		Normal, limited back to stationary	Alluvial, paludal, lacustrine, evaporites	
	Back-arc compression		Vertical upthrust, basement involvement, stationary	Alluvial, paludal, lacustrine	
	Forearc		Normal, limited step-back (?)	Paralic, pelagic, submarine fan	
STRIKE-SLIP	Transtension (pull-apart)		Strike-slip and normal, variable motion depending on the side of the basin. One side stationary, two sides step-back	Early-pelagic and submarine fan. Late-lacustrine	Low to moderate thermal activity except under conditions of extreme transtension. Rapid, deep (1000's m) burial hastens diagenesis.
	Transpression		Low-angle reverse	Alluvial plain	Very low thermal activity, shallow burial, and little early diagenetic modifciation.

DIVERGENT	Proto-rift	Normal. Complex, repetitive step-back. Basin partitioning. Mature development	Evaporites, lacustrine, and alluvial facies	High thermal activity. Rapid subsidence (4-7 km/10 m.y.) Extensive diagenesis including zeolitization and feldspathization and addition of carbonate and quartz cement. Deep burial under shelves destroys virtually all porosity.
	Intracratonic	Normal. Complex, repetitive step-back. Immature development	Evaporites, lacustrine, and alluvial facies	
	Aulacogen	Normal. Complex, repetitive step-back. Immature development	Paralic, shelf, pelagic and submarine-fan facies	

Table 1.1 Tectonic settings for alluvial fans

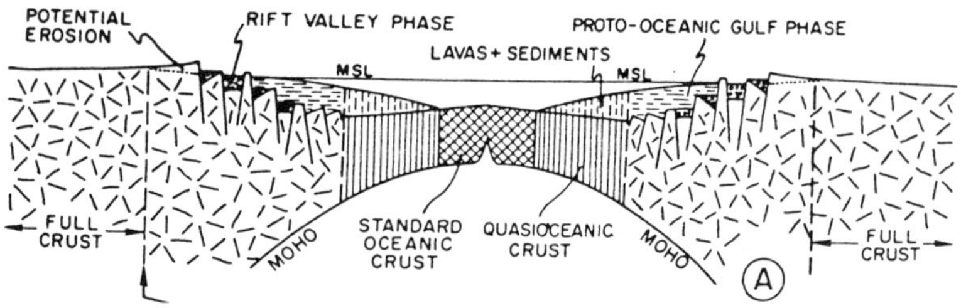

Figure 1.2 Alluvial fan deposits associated with a mature extensional tectonic setting. The alluvial fan deposits can be intercalated with lavas and evaporites and can be spatially discontinuous because of basin partitioning associated with horsts and rotation of fault blocks (from Dickinson, 1976).

modification of primary porosity will be the rule rather than the exception.

Crustal convergence manifests itself most notably in the development of low-angle reverse faults typical of foreland fold-thrust belts. However, foreland basins are not likely settings for thick alluvial-fan sequences, except along their structurally mobile margins. Partly, this is because the low-angle reverse faults, with relatively non-resistant sedimentary rocks making up the allocthanous plates, do not provide a sufficient long-term setting for abundant sediment supply and abrupt gradient change between the frontal thrust and alluvial plain. Also, alluvial fan facies that do form have a large probability of being cannibalized by a subsequent thrust. Instead, coarse alluvial facies in foreland basins are better understood in the framework of broad alluvial plain or apron models. The same models also apply probably to understanding thick alluvial deposits associated with continental suture zones where high angle faults are unlikely along basin margins.

Nevertheless, high-angle basin-bounding faults are present in zones of crustal convergence. They are complex in terms of setting and origin. For example, the inner boundary of a forearc basin is normal and high-angle,

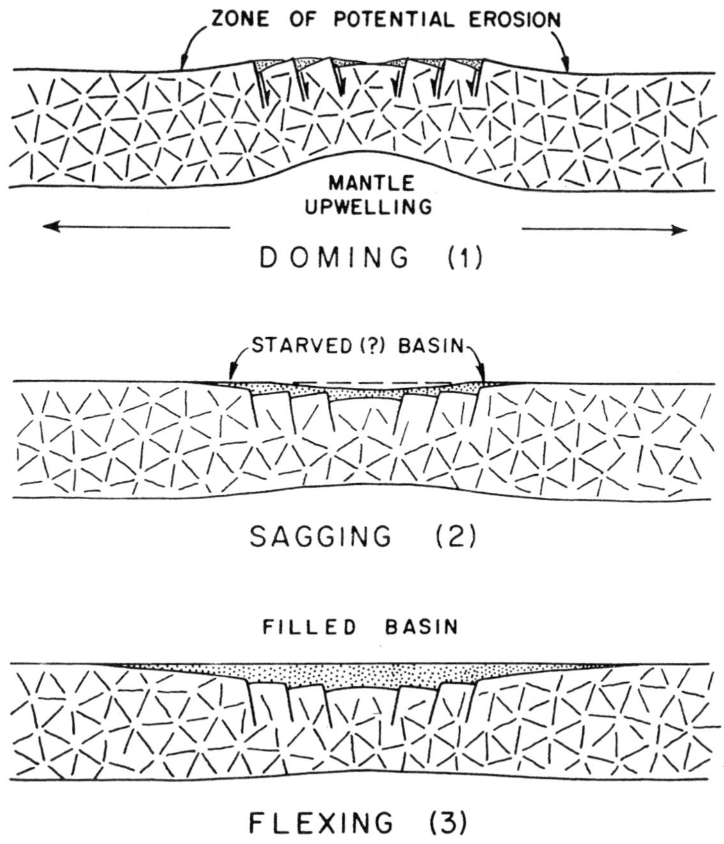

Figure 1.3 Alluvial-fan deposits associated with immature crustal divergence in an intraplate setting. Although clastic deposition occurs during all three stages, alluvial fan deposition is restricted only to stage 1 (doming), (from Dickinson, 1976).

(Dickinson, 1974, Fig. 10) and is a likely setting for fan-delta formation (Fig. 1.5). Local volcano-tectonic subsidence or extension over a thermal welt could lead to high-angle normal faulting of limited step-back nature within the magmatic arc (Fig. 1.5). The lower Cretaceous alluvial-fan and fan-delta deposits of northern Chile were deposited in this type of intra-arc basin (Flint, *et al*, 1986). Also, superimposed on the relatively recently defined metamorphic core-complexes within or adjacent to the

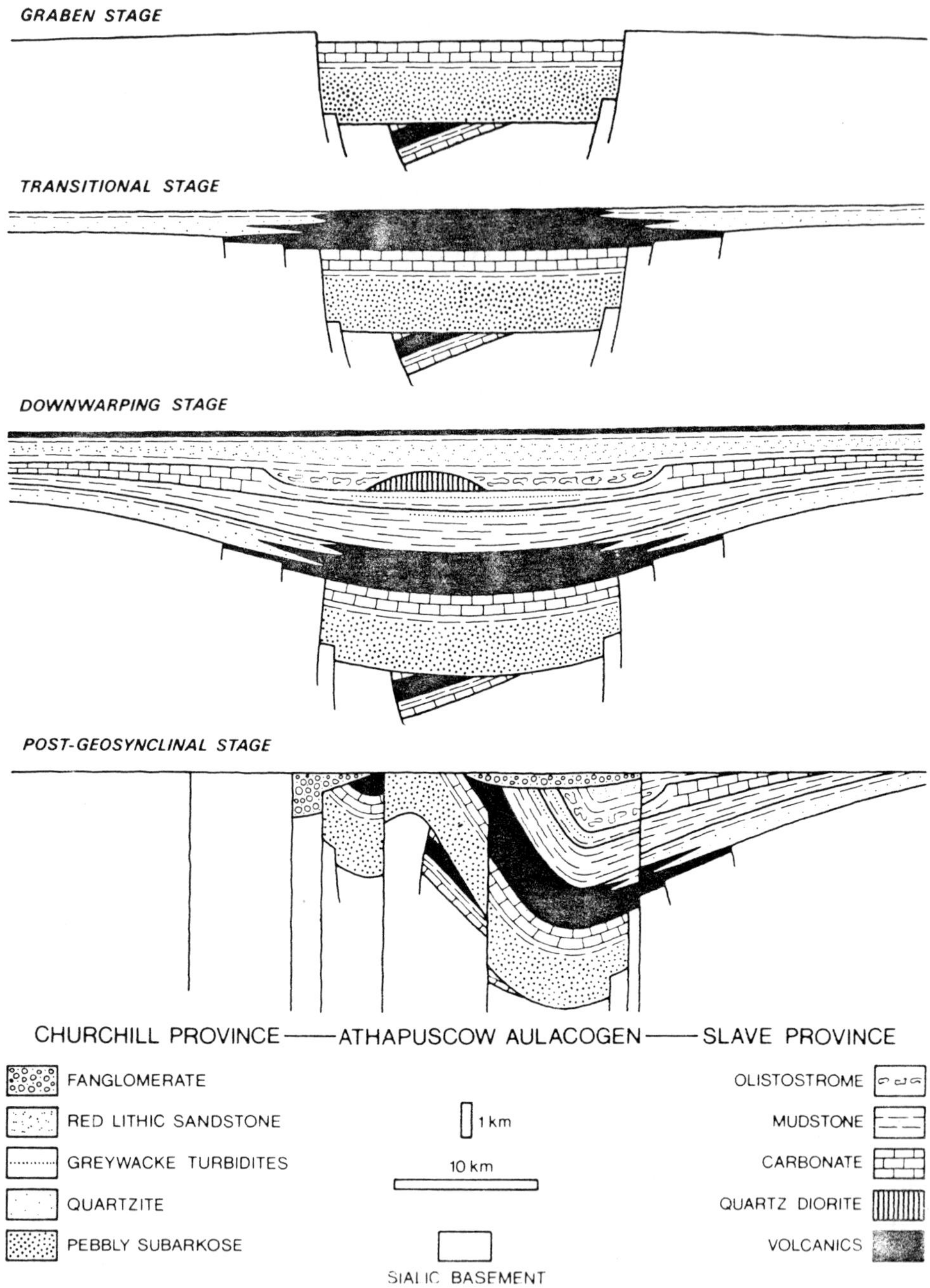

Figure 1.4 Stages in the development of an aulacogen. Alluvial-fan deposits can accumulate early in the aulacogen's history (pebbly subarkose) and also later associated with reactivation of basement faults (fan conglomerate) (from Hoffman et al, 1974).

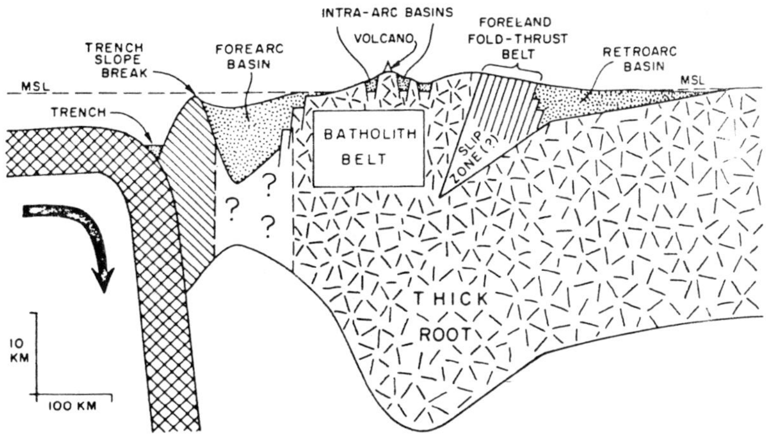

Figure 1.5 Sites of alluvial-fan formation in an area of crustal convergence. Fan deposits can form in the forearc and intra-arc basins. Alluvial plain deposition probably characterizes the retroarc basin (from Dickinson, 1976).

Cordilleran magmatic arc are thick alluvial fan deposits that accumulated in basins formed by distension and bounded by normal faults of step-back character (Gager, 1984; Grover, 1984).

In the back-arc setting, back-arc spreading can lead to block faulting of Basin and Range style. Also, back-arc compression can generate high-angle vertical upthrust structures involving basement rocks, characteristic of the so-called Wyoming structural province (Fig. 1.6). Extensive fanconglomerates along the east flank of the Big Horn Range in Wyoming, containing clasts up to 7 m in diameter, are deposits of alluvial fans associated with the basement thrusts (Sharp, 1948). Both the Basin and Range type faults and the vertical upthrusts tend to possess a relatively permanent location, that is, they do not back-step.

Alluvial fans in sedimentary basins formed in zones of plate convergence seldom will show down-dip connection with marine facies. An important exception would be alluvial fans in forearc settings. Generalization about diagenetic patterns is nearly impossible because of the extremely variable thermal regime in and adjacent to magmatic arcs. However, typically detritus is immature, rich in

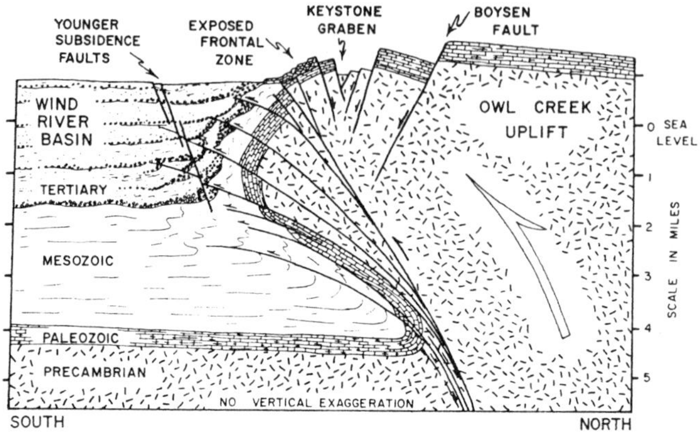

Figure 1.6 Alluvial fan deposits associated with basement thrust faults in the Wyoming structural province, (from Wise, 1963).

volcanoclastic debris, and therefore, strongly susceptible to cementation.

Sedimentary basins associated with strike-slip motion are among the most complex in origin and evolution. They may develop along plate margins or some distance from plate boundaries. They may terminate at spreading centers or in subduction zones, and involve either continental or oceanic crust. Characteristically, major strike-slip fault systems are curvilinear or braided. Thus, horizontal movement causes local zones of extension and compression (Fig. 1.7). Zones of extension, more properly referred to as transtension because of a component of transcurrent movement, are potential sites of deep sedimentary basins bounded by high angle faults. Thus these so-called pull-apart basins are potential sites of alluvial fan sedimentation. Zones of compression or transpression are marked by thrust faulting, folding, and uplift, and are logical settings for alluvial plain deposition, for the same reasons alluvial plains are expected in foreland basins.

In zones of transtension the dominant strike-slip fault is stationary in location. But, both the parallel subordinate high-angle fault on the opposite side of the pull-apart basin and the bounding fault at the head of the basin, which is oriented at a high angle to the main vector of strike-slip motion, are repetitive step-back faults (Fig. 1.8). Consequently, stacked sequences of alluvial fan facies associated with the various faults will differ.

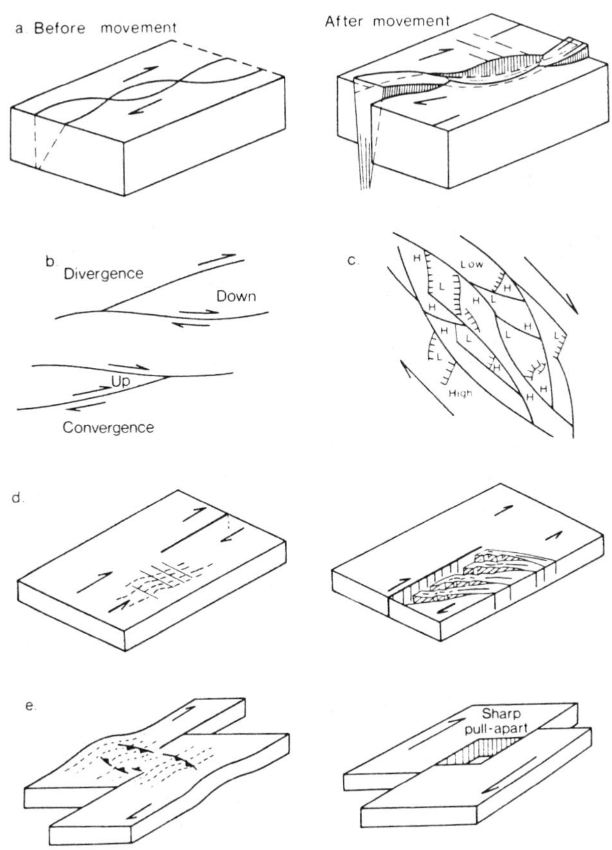

Figure 1.7 Types of strike-slip fault patterns that produce extensional basins and compressional uplifts (from Reading, 1980).

Further complicating interpretation of the fan facies is that many pull-apart basins are less than 10 km across, implying close juxtaposition of fan and non-fan facies. Non-fan facies may include both lacustrine and marine basin facies. In the California continental borderland basins, pelagic and hemipalagic facies and submarine fan and shelf facies are closely associated with the alluvial fan rocks. Elsewhere, as is the Yallahs Basin of Jamaica (Wescott and Ethridge, 1980), carbonate bank deposits are found.

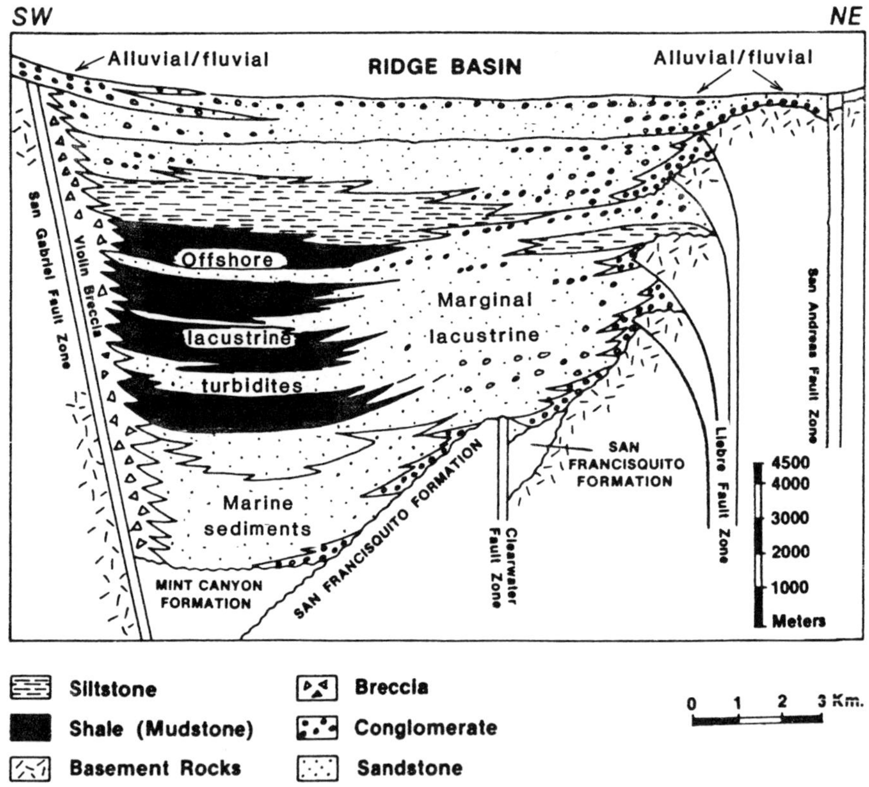

Figure 1.8 Alluvial fan deposits in the Pliocene Ridge Basin of California. Notice difference in geometry of alluvial fan deposits from one side of the basin to the other (from Mitchell and Reading, 1978; Link and Osborne, 1982)

Igneous activity in most strike-slip settings generally is sparse. Extreme transtension however, can bring about deep crustal fracturing and magmatism, but this is rare in continental settings. Nevertheless, heat flow in extensional settings often is higher than normal which is important for source-rock maturation of hydrocarbons. Given the common close juxtaposition of organic-rich sediments and porous alluvial-fan sands, and the great potential for both stratigraphic traps (e.g. unconformites) and structural traps (e.g. drag folds associated with basin-margin faults), basins in strike-slip settings are prime sites of petroleum production as so amply demonstrated in the Tertiary of western California (Blake et al 1978).

2

Alluvial-Fan Morphology

INTRODUCTION

Optimum reservoir prediction is dependent on knowledge of geometry, dimensions, and orientation of potential reservoir bodies. In alluvial fan complexes, these properties relate directly to the external and internal fan morphology. Although the external morphology of fans is easily studied and, therefore, relatively well understood, the internal structure of modern fans normally is poorly exposed and, therefore, not well understood. Consequently, our knowledge of the internal structure of fans largely is based on examination of ancient fans. A potential weakness in facies models of alluvial fans results from inadequate and over-simplified attempts to make a genetic association between the external morphology of modern-day fans and the internal facies distribution of fans as observed in the ancient.

EXTERNAL MORPHOLOGY OF ALLUVIAL FANS

Fans can be of two types: wet or humid fans and dry or arid fans. Distinction between the types of fans is based on shape, depositional process, and fundamental difference in facies patterns. Because of the latter, knowledge of the climatic setting of a particular fan complex is critical to understanding its facies architecture.

Wet fans are large (1000's km^2) low gradient (m's/km) landforms dominated by perennial fluvial processes. Systematic lateral migration of a single or limited number of major trunk streams on wet fans tends to give rise to sheet-like geometry of sand/gravel bodies with relatively sharp vertical segregation by overbank deposits containing well-developed paleosols. Down-fan systematic decrease in grain-size is common. In contrast, dry fans are smaller (100's km^2) and higher gradient (10's m/km) and are constructed by fluvial and sediment gravity processes. Complex migration and abandonment of feeder channels gives rise to complex facies geometry and stacking and complex longitudinal variations in grain size. Vertical facies segregation often is weakly developed and mature soil profiles are uncommon. Processes of wet-fan sedimentation can and have been studied in pro-glacial environments of arctic regions (Boothroyd and Ashley, 1975). However, the small scale of wet-fans in such regions and unusual dynamics of glacial meltwater flow associated with abnormal hydrostatic head renders it difficult to apply directly the results of these studies to the giant wet fans of the past.

The Kosi River fan, developed off the south flank of the Himalayas in Nepal and India, is the most widely cited example of a typical wet fan (Gole and Chitale, 1966). The Kosi fan is about 200 km in radial length. It has been built largely by deposition from the Kosi River, a braided river which ranges from 6-15 kms wide with a gradient of about one m/km. The Kosi River has shifted laterally 110 km in the last 250 years resulting in a sheetlike accumulation of sand and gravel. Conglomerate is restricted to the upper 15 km of the main channel.

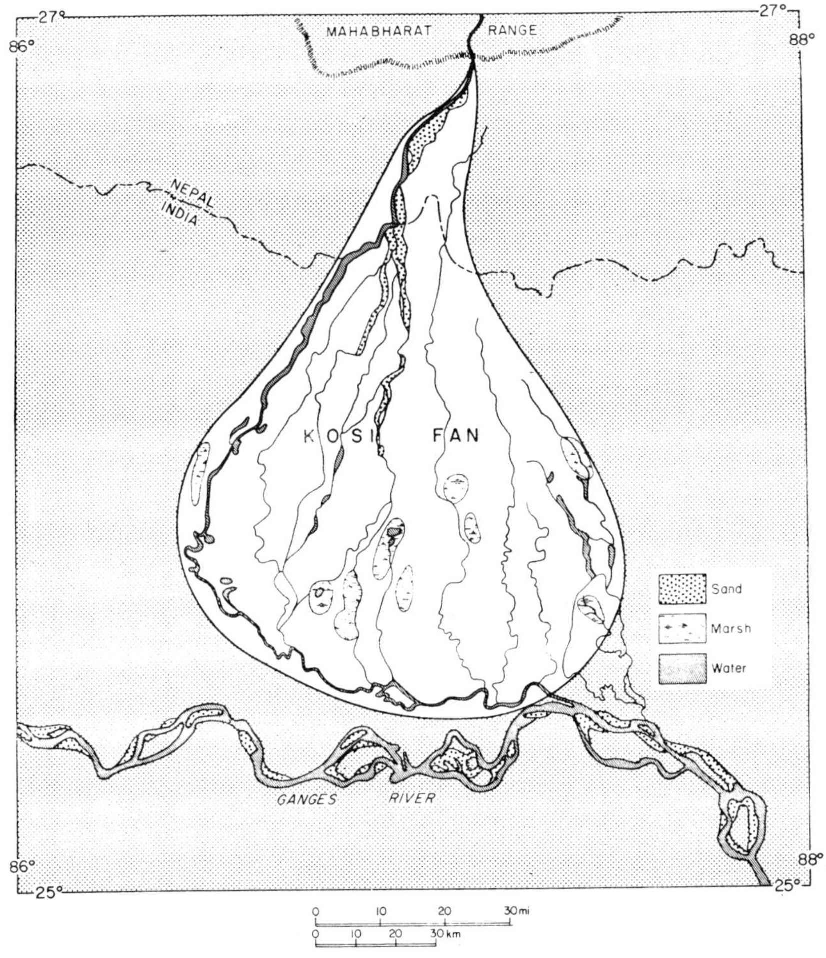

Figure 2.1 Map of Kosi fan in Nepal and India (from Seni, 1980).

Perhaps the best facies models for ancient wet fan deposits are those based on alluvial-plains and, in fact, most of the well-documented ancient wet-fan deposits (e.g.

Cambrian Van Horn Sandstone of Texas), probably are best understood within a framework of alluvial plain concepts and models. Unfortunately, few regional studies of ancient alluvial-plain complexes exist. One notable example is Seni's (1980) analysis of the Neogene Ogallala Formation in Texas. Seni examined over 15,000 well logs in putting together this benchmark study of an ancient alluvial plain which developed off the Rocky Mountains and extends from the Dakotas south into Texas (Fig. 2.2). Predictably, the alluvial plain contains sand and gravel sheets up to 200 kms wide. Dip oriented sand-rich belts bifurcate down-dip and range from 10 to 50 kms. wide (Fig. 2.3). An important lesson that can be gained from careful examination of the architecture of the alluvial plain complex in the Ogallala, is that application of wet-fan or alluvial plain models to dry fans, or vice versa, is, at best, dangerous.

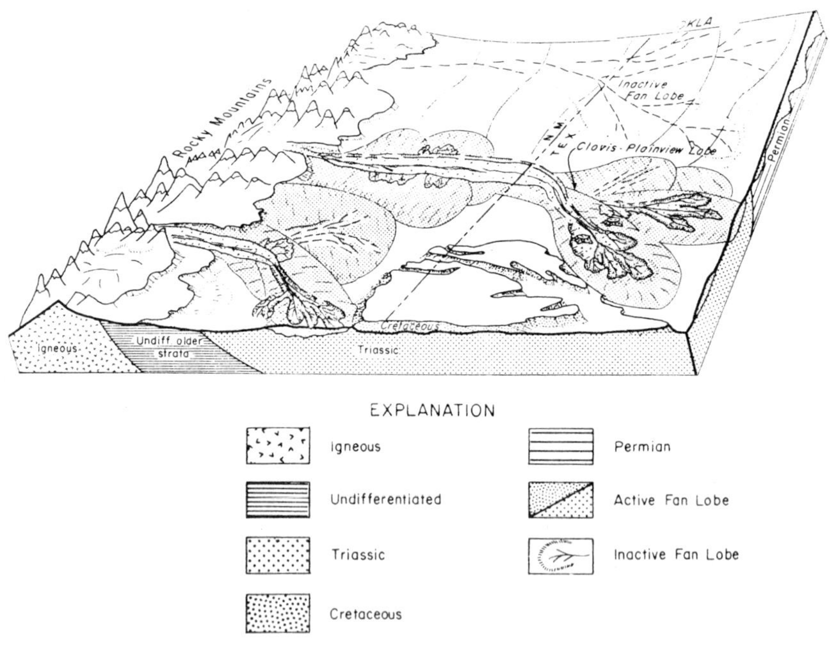

Figure 2.2 Schematic block diagram of a part of the Ogallala alluvial plain system (from Seni, 1980).

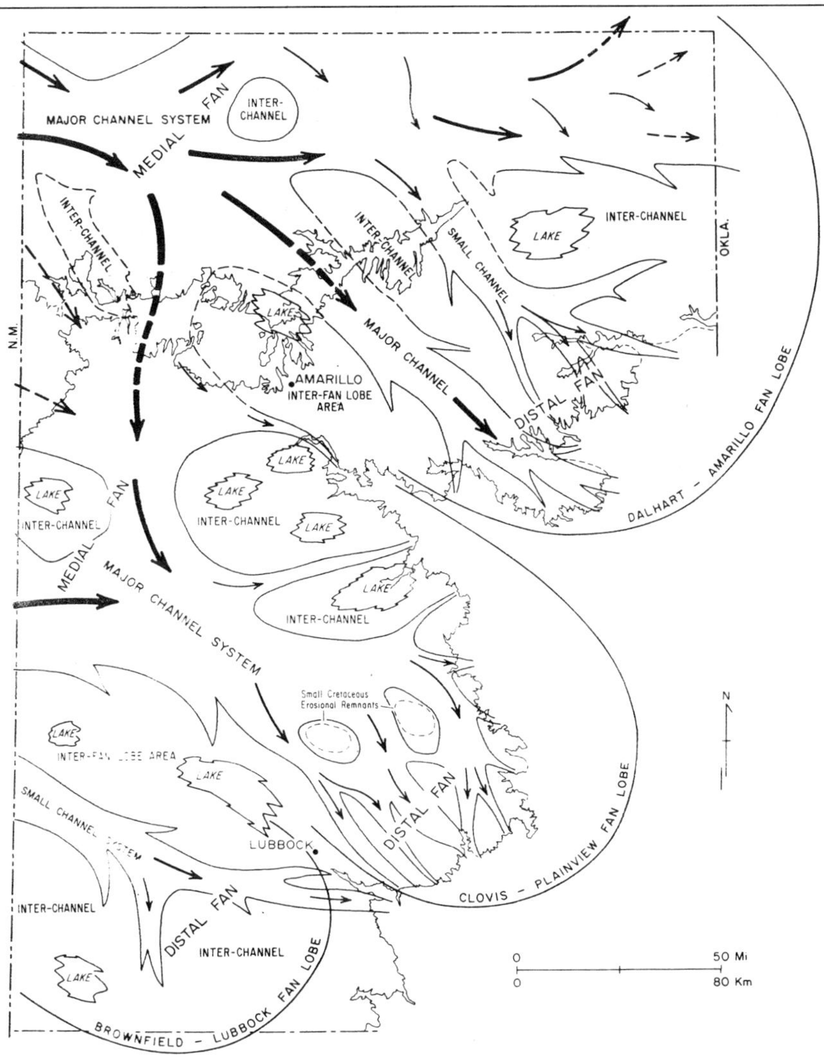

Figure 2.3 Schematic illustration of depositional facies and sediment dispersal systems within the Ogallala Formation in Texas (from Seni, 1980).

MORPHOLOGIC ZONATION OF ARID ALLUVIAL FANS

Arid alluvial fans, both modern and ancient, have been the focus of more detailed study than wet fans. A product largely of the study of ancient fans is the facies zonation of arid-fans into inner-fan (proximal or fan-head), midfan, and distal fan. (Figs. 2.4, 2.5). The zonation in ancient fans is largely based on texture, although implicit is a process and micro-environment basis as well.

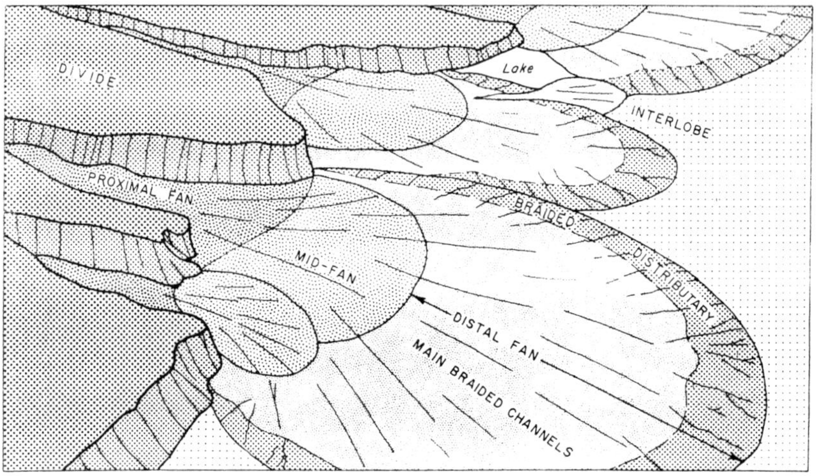

Figure 2.4 Schematic representation of morphologic zonation of an ancient alluvial fan system--the Cambrian Van Horn Sandstone, Texas (from McGowen and Groat, 1971).

The dominant feature of the inner fan is the fan-head trench, which can be tens of meters deep and may or may not be bounded by natural levees. Streamflood, sediment gravity flow deposition (e.g. debris flow, landslide) and sieve deposition are characteristic processes of the inner fan, and all can result in accumulation of conglomerate. Coarse fanhead trench deposits of the inner fan may be separated by broad areas of finer-grained interchannel sediments.

The mid-fan area begins beyond the intersection of the fan-head trench with the fan surface. Literally hundreds of shallow braided distributary channels may cross the mid-fan which is dominated, as a consequence, by stream deposition. In contrast to the conglomeratic fan-head, the mid-fan is sand/gravel dominated. Sand bodies are highly discontinuous laterally but semi-continuous along radial transects.

The distal fan processes and facies intimately associate with processes and facies of the fan fringe environments and separation of the two often is difficult. The distal fan is characterized by non-channelized sheetflood deposition. Sheet-like deposits of sand and finer clastics are the most abundant facies. A limited number of braided channels may also extend across the distal fan.

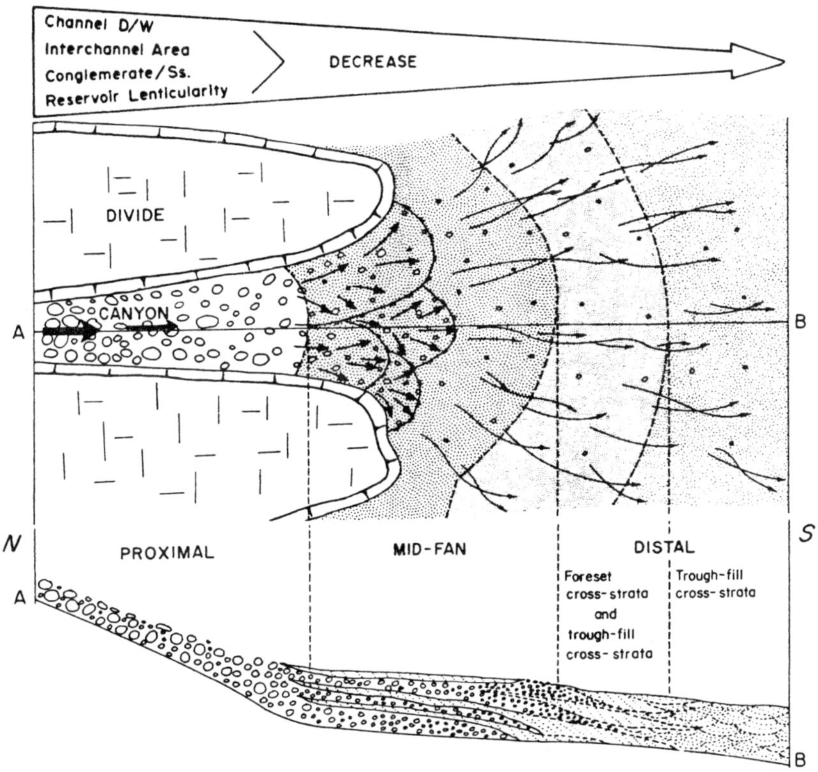

Figure 2.5 Schematic cross-section and plan view of an ancient alluvial fan illustrating process/facies variability as a function of morphologic zonation (after McGowen and Groat, 1971).

Two widely accepted implications derive from the facies zonation of arid alluvial fans. First, is the implied existence of a systematic down-fan decrease in grain size, or at least conglomerate to sandstone ratio. Secondly, the zonation, if it exists, would imply that fan progradation should superimpose inner fan facies on mid-fan and distal fan facies to give rise to coarsening upward vertical profiles; fan retrogradation should produce the opposite trend. The external morphology of modern fans suggests that such simple patterns are unlikely to be present and that facies zonation of arid fans may be an over- simplified attempt to model ancient fans. For example, Bull (1968) defined two end members of fan morphology (Fig. 2.6). In example A in Figure 2.6, deposition takes place near the mountain front, no fanhead trench exists, and little fan dissection occurs. In example B in Figure 2.6, deposition is concentrated near the toe of the fan and the sediment is delivered to the toe via a long trench. The causes of fan-head entrenchment are widely debated and certainly relate to tectonic controls on base-level changes, as well as climatic induced effects which we will discuss in more detail later. The existence of both types of fans within the same block-fault valley, as reported by Schumn (1977) for Death Valley,

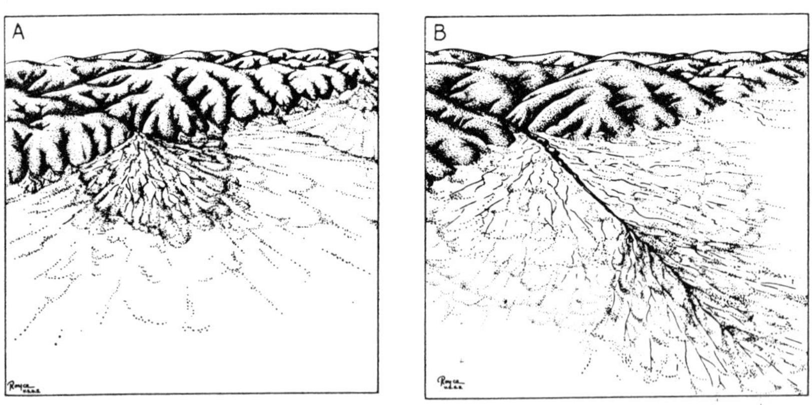

Figure 2.6 Contrasting styles of alluvial fans. (A) Area of deposition adjacent to mountain front; (B) area of deposition downfan from the mountain front due to fan head entrenchment (from Bull, 1968).

suggests very localized causes. Regardless, the spectrum of fan morphologies poses profound sedimentologic implications insofar as the facies zonation of fans is concerned. For example, given end member B, a systematic downfan decrease in grain size and ratio of conglomerate to sandstone would be unlikely.

Simple relationships affecting prediction of geometry of reservoir bodies also breakdown as revealed by a comparison of the distribution of gravels of different ages on the Death Valley fans shown in Fig. 2.7. In the Johnson County fan, unidirectional migration of a single or small number of major channels has given rise to a sheet geometry for the gravel. In contrast, on the Trail Canyon fan, multiple major drainages manifest themselves in the development of more lenticular gravel bodies oriented parallel to fan radii. An even more complicated geometry of gravel deposits occur on the Hanaupah Canyon fan. In none of the examples does the gravel distribution fit the pattern of simple facies zonation so widely applied to ancient fan deposits.

Even more disconcerting is the potential for incredibly complex vertical stacking of facies, if an alluvial fan complex evolves like the one shown in Figure 2.8. Later we will generate some ideal thick vertical profiles for alluvial fans and consider causes of variability in the profiles. Suffice it to say at this point that predictive models for vertical stacking and lateral juxtaposition of fan facies cannot possibly be viable if they are as simple as the typical examples widely referred to in existing literature.

FAN FRINGE MORPHOLOGY

Introduction

Anstey (1965) concluded that the overwhelming number of

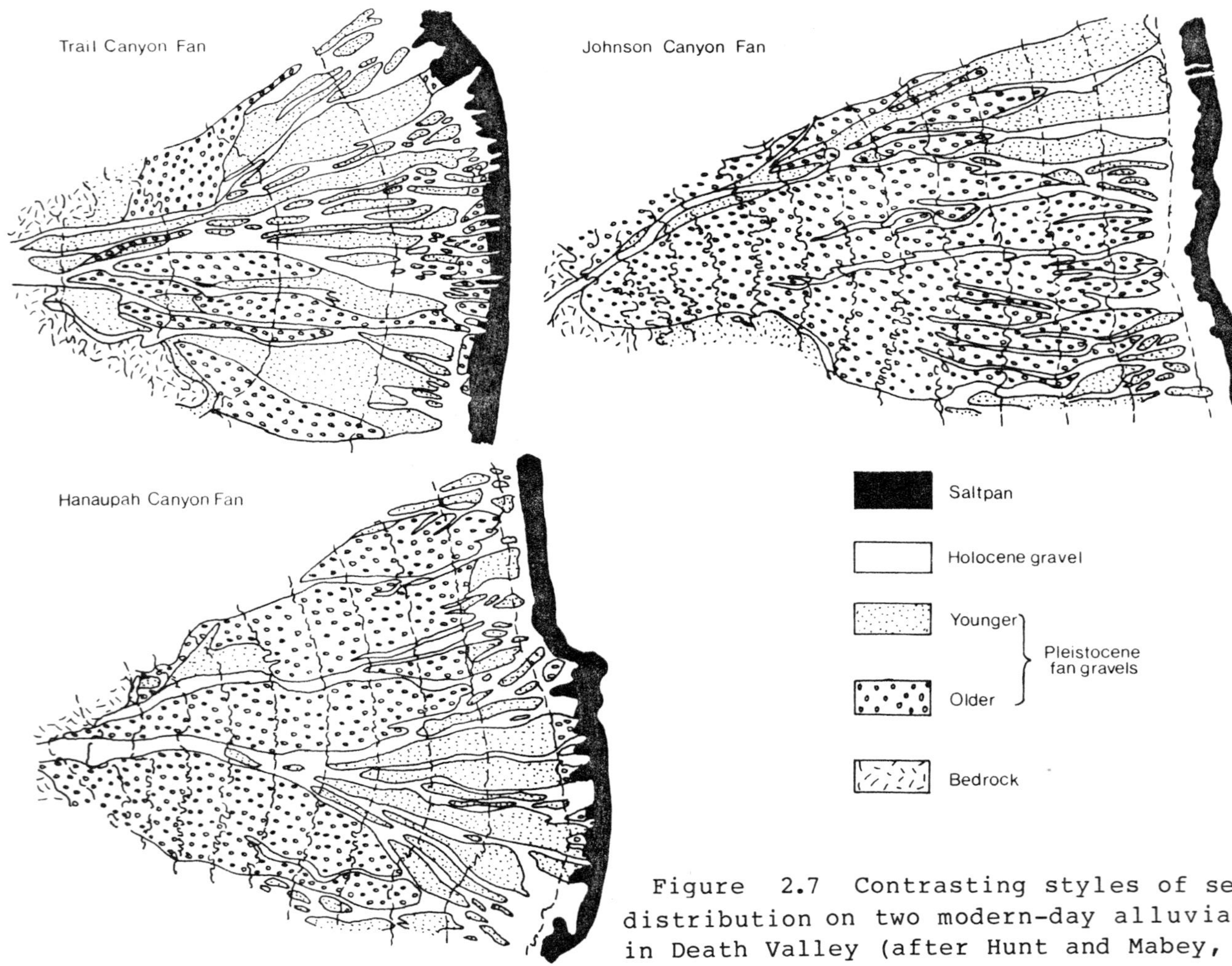

Figure 2.7 Contrasting styles of sediment distribution on two modern-day alluvial fans in Death Valley (after Hunt and Mabey, 1966).

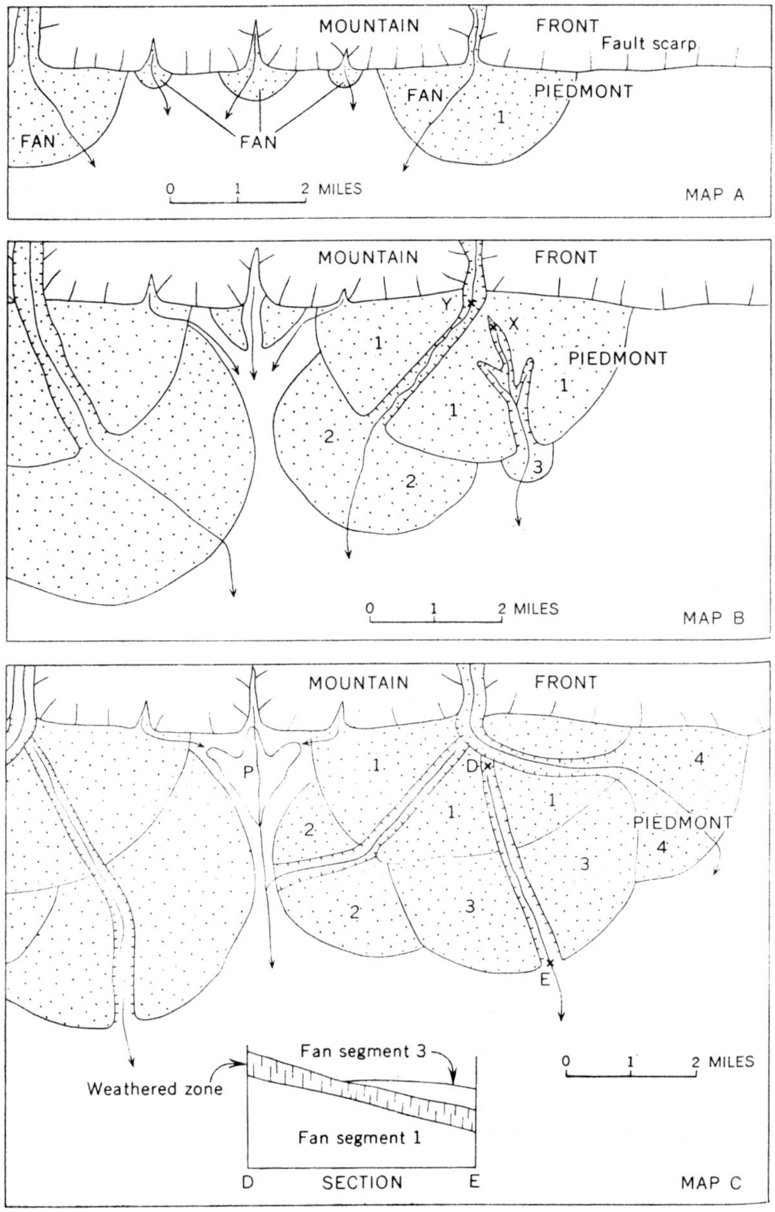

Figure 2.8 Maps showing evolution of alluvial fans. This evolution would result in complex vertical stacking sequences of alluvial-fan facies (from Denny, 1967).

modern arid fans have radii between about 2 and 8 kms. This is an important observation because it means that facies belts of alluvial fan origin which are much wider than 8 kms, and which are located in tectonic settings of limited back-step faulting, almost certainly will contain facies of fan-fringe environments. The texture, geometry and orientation of these fringe deposits likely will be in sharp contrast to the same properties of facies making up the fan proper. This is another source of complication, as we illustrate later, in doing vertical profile facies analyses in ancient fans. And it underscores the need for future study of the vertical and lateral arrangement of intercalated fan and fan-fringe facies, including playas and lacustrine facies of the type recently published by Flint (1985) and Mancini, et al. (1985).

Trunk Alluvial Systems

Alluvial-fan fringes most commonly merge with alluvial deposits of rivers flowing longitudinally down a valley at a high angle to the out-building fan (so-called trunk, basin-axial or syntaxial streams) (Fig. 2.9). These axial rivers can be meandering, braided, or anastomosed. They can be transporting sediments coarser than or finer than the distal-fan sediments. The rivers may truncate the fan or rework fan deposits. Eolian dune sands may or may not be present in interchannel areas. At the very least, reservoir bodies in the axial alluvial deposits will be oriented at high angles to the potential reservoirs on the fan itself.

Fan Deltas

Fan deltas form where the alluvial fan progrades into a standing body of water from an adjacent highland (Holmes, 1965). Therefore, fan-delta systems are composed of subaerial alluvial-fan facies, transitional, and submarine

Figure 2.9 Aerial view of alluvial fans in Death Valley, California showing a variety of alluvial systems in areas fringing the fans.

facies. Recognition of this association of facies in close vertical and lateral contact is the basis for recognition of ancient fan-delta complexes and will be descussed in more detail in the next chapter.

The sedimentology of fan deltas, particularly the subaqueous parts of the system, is at the cutting-edge of studies of modern and ancient depositional systems. Undoubtedly, this is because the importance of fan-delta deposits as reservoirs for oil and gas is just now being fully recognized (Ethridge and Wescott, 1984). Stratigraphic and structural traps abound; lateral connection of potential reservoir beds with hydrocarbon source beds of marine pelagic and hemipelagic origin is well developed.

Clearly the nature, orientation and distribution of the potential shoreline and offshore reservoirs in fan deltas is a function of their morphology, and their morphology is a function of their setting. Ethridge and Wescott (1984) tentatively identified three fan-delta models. Slope fan-deltas, which form in convergent and transcurrent tectonic settings, prograde across very narrow shelves down relatively steep slopes. The subaqueous part of these complexes will consist of well-sorted littoral sand bodies oriented parallel to the shoreline which grade seaward into slope deposits of mud and submarine fans (Fig 2.10, 2.11). The second type of fan delta develops where broader shelves exist (e.g. in intracratonic basins and divergent plate-tectonic settings) and is characterized by extensive shelf progradational sequences,(Fig. 2.12).

Numerous variants of the shelf model exist, because of the local relative importance of tidal, wave and fluvial induced shoreline processes. As a consequence, and as Kleinspehn et al (1984) suggested, shelf fan-deltas can be classified similarly to normal fluvial deltaic shorelines. This is a powerful idea because it enables clearer understanding of the morphology and sedimentologic processes of shelf fan-deltas and, in turn, better prediction of reservoir quality, geometry and distribution. For example, wave-dominated fan-deltas will contain a large proportion of well-sorted, shoreline-parallel sand bodies (e.g. beach, foreshore, shoreface, barrier islands). Tide-dominated fan

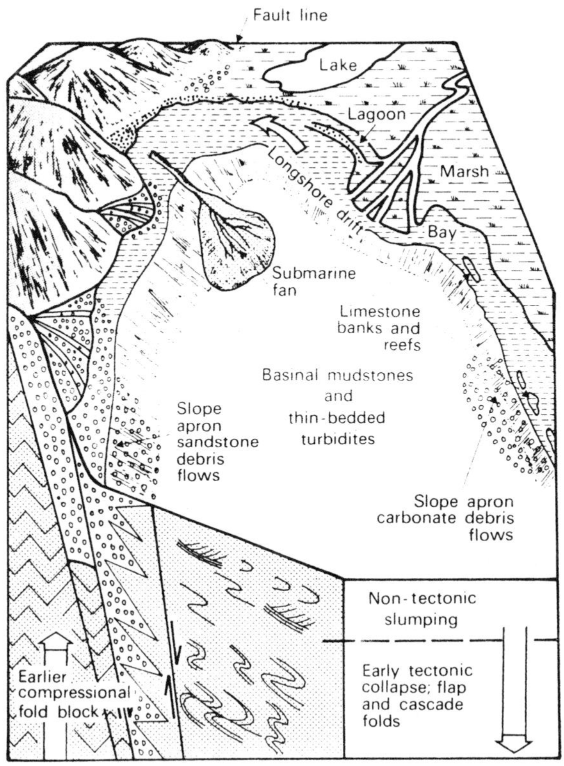

Figure 2.10 Typical slope-fan-delta showing close spatial association of subaerial and subaqueous fan deposits (from Mitchell and Reading, 1978).

deltas should contain shoreline perpendicular sand bodies (e.g. tidal inlets, subtidal bars and channels) and possibly offshore carbonate banks and muds. River-dominated fan deltas on broad shelves will contain bar-mouth sands oriented perpendicular to the shoreline. However, because avulsion is a common process on subaerial alluvial fans, frequent shifting of the locus of braided stream discharge

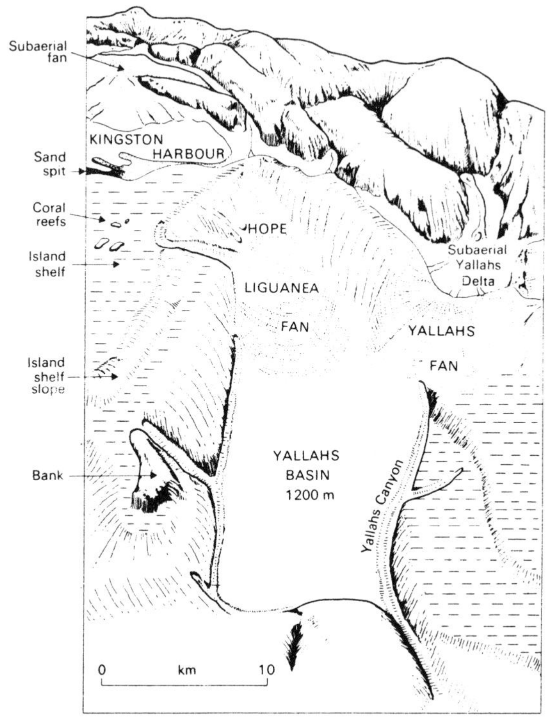

Figure 2.11 Sketch map of the Yallahs basin, Jamaica showing close spatial association of subaerial and subaqueous fan deposits (from Mitchell and Reading, 1978)

on the deltaic plain will permit reworking of sand by shoreline processes. Therefore, bar mouth sands oriented perpendicular to the shoreline may be intimately associated with wave-reworked sands oriented parallel to the shoreline as observed in the ancient by Kleinspehn _et al_ (1984) in the Carboniferous-Permian in Spitsbergen and in the Holocene by Hayward (1985) in the Gulf of Aqaba. Moreover, evolution of coastlines through time might superimpose one type of shelf fan-delta complex on another as illustrated by Roe and Steel (1985). Also, in the Fountain Formation (Pennsylvanian) near Colorado Springs the oldest part of a fan-delta complex shows tidal influence, but in the upper part evidence for influence by only wave/river mouth processes exists

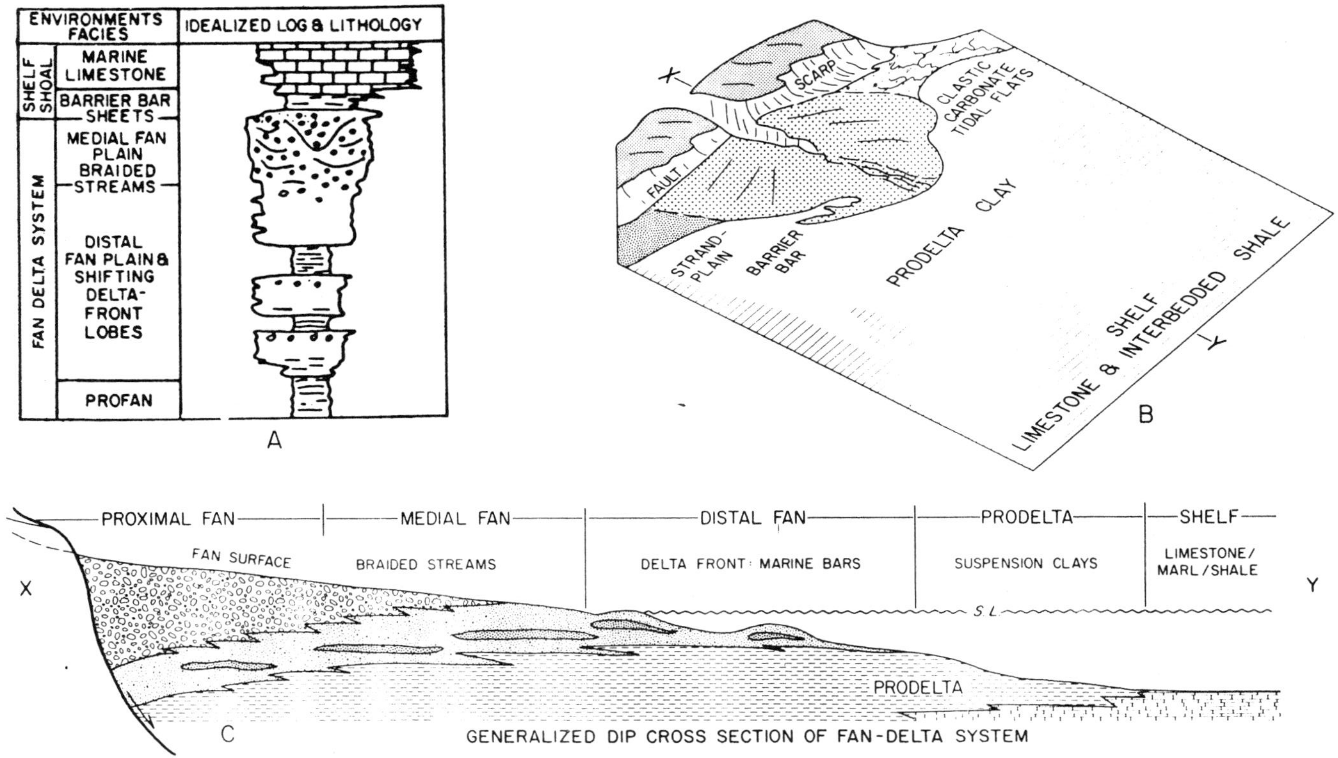

Figure 2.12 Facies and morphology of a typical shelf fan-delta (from Brown, 1979).

(Suttner, et al 1985). In this case, the oldest part of the fan delta accumulated on a shelf with relief related to the underlying Mississippian-Pennsylvanian unconformity. Local topographic highs funneled and intensified tidal currents and, at the same time, dampened the effects of waves. As the surface of relief was gradually buried, tidal dominance gave way to wave dominance in the younger part of the fan delta.

An important difference between the facies geometry of slope and shelf fan-deltas derives from the difference in

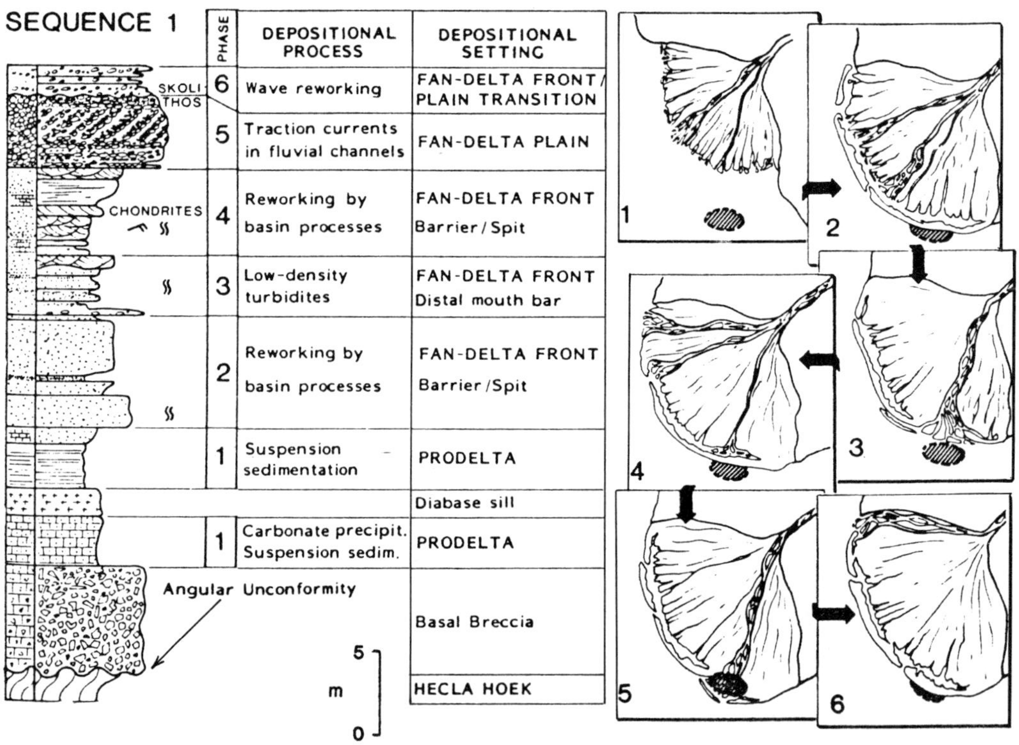

Figure 2.13 Evolution of a fan-delta facies sequence. Note how lateral migration of the trunk stream through time causes a superposition of shoreline-parallel sand bodies (barrier-spit) and shoreline-perpendicular sand bodies (distal mouth-bar) (from Kleinspehn, 1984).

seafloor gradients in the two settings. In slope fan-deltas, because of the steep sea-floor gradient, minor changes in sea level normally do not cause areally extensive transgression and regression. Consequently, the zone of maximum stratigraphic intercalation of marine and nonmarine facies is narrow. In contrast, in shelf fan-deltas, even minor changes in sea level cause areally extensive transgression and regression. Interbedding of marine and nonmarine facies, therefore, is common and the zone of interbedding is broad (Figs. 2.14, 2.15). Moreover, because of the higher preservation probability of progradational

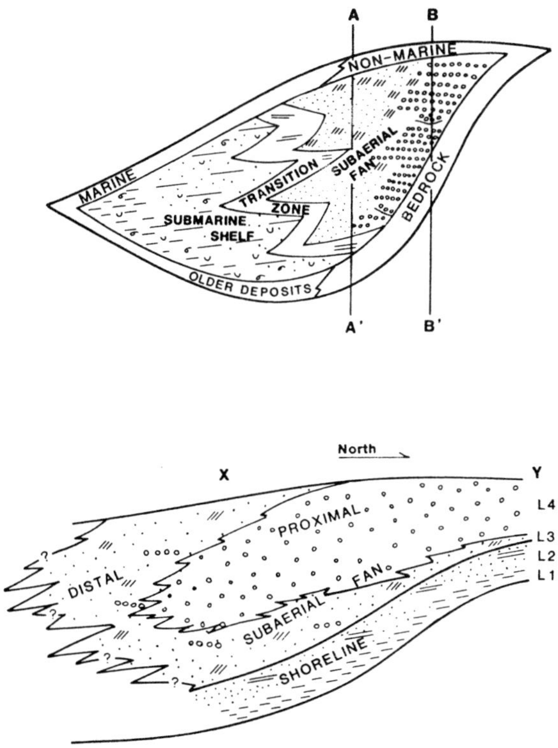

Figure 2.14 Schematic representation of difference in width of zone of marine and non-marine facies mixing in slope (left) and shelf (right) fan-deltas (from Melvin and Knight, 1984).

sequences, relative to retrogradational sequences, stacked nonmarine to marine facies sequences are probably common in shelf fan-deltas.

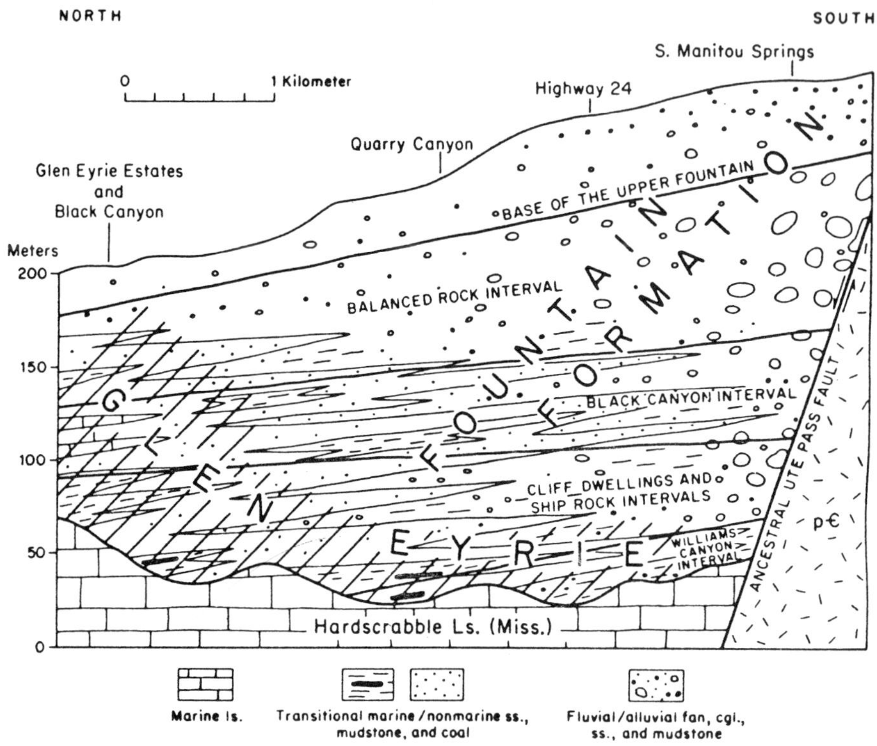

Figure 2.15 Schematic cross-section of the Fountain fan-delta near Colorado Springs. The Fountain fan-delta is a shelf fan-delta. The zone of marine/non-marine facies mixing is relatively wide and stratigraphic transitions from marine to non-marine facies (i.e. progradational sequences) are common (from Suttner <u>et</u> <u>al</u> in press).

The third type of fan-delta recognized by Ethridge and Wescott (1984) is the Gilbert-type model. This type of fan delta builds into shallow, probably highly restricted standing bodies of water through progradation of large-scale (2-5 m thick) gravelly or sandy foresets. An excellent description of a Pliocene Gilbert-type fan delta in the Vera basin in southeastern Spain is given by Postma and Roep (1985). We suspect that with future study the Gilbert-type fan-delta deposit will be integrated as a separate facies in some variant of the shelf model.

Although lacustrine fan-deltas are indirectly included in the Ethridge and Wescott classification scheme, their unusual setting, size and association of facies, we think should set them apart.

3

Processes, Facies, and Facies Distribution on Alluvial Fans

INTRODUCTION

A limited number of depositional processes are active on alluvial fans and they yield a small finite number of deposits. Stream flow, both in the form of channelized and unchannelized sheetflood, is the dominant, if not nearly exclusive depositional process on wet fans and also accounts for the largest volume overall of deposits on arid-fans. Sediment gravity flows, including debris flows, mud flows and landslides are important processes on arid fans, but rarely observed on wet fans. Sieve and eolian processes of deposition are subordinate.

In a very general way, ancient alluvial fans can be zoned on the basis of the relative proportion of the deposits of the above processes. The fanhead or proximal fan contains by far the largest volume of gravity flow and sieve deposits. The mid-fan is dominated by braided channel flow, and the outer or distal fan by sheetflow and, in some settings, eolian deposition. But, such zonation must be done cautiously, as we emphasized in the last chapter. This is because complicated vertical and lateral juxtaposition of the different deposits can result from morphologic evolution of the drainage basin, scarp retreat, unusual climatic events, lateral migration of the locus of fan deposition, either gradually or abruptly through avulsion, and changes in base level. These complications are discussed in detail in Chapters 4 and 5. At the very least, however, the association of facies which are products of the fan depositional processes is unique. Recognition of the elements of the facies association permits rather easy and confident recognition of alluvial fan systems in the rock record. In this chapter we focus on the diagnostic criteria of these individual alluvial fan facies.

BACKGROUND

Conglomerate Facies Description

As a result of the high gradients of alluvial fans, conglomerate is an important component of fans. Distinction between conglomerates of various origin is dependent on systematic rigorous description. Figure 3.1 is a guide to the critical observation one must make in inferring the origin of conglomerate. Although not directly implied by Figure 3.1, estimation of maximum clast size (usually determined by averaging the long diameters of the 10 largest clasts in an outcrop of some given minimum area) and thickness of the conglomerate bed also are important observations.

Flow Regime and Bedform Theory

Streams, flowing both confined and unconfined on alluvial fans, behave as Newtonian fluids (i.e. they possess little strength or resistance to shear). Therefore, flow regime and bedform theory is applicable to understanding the deposits of streamflow on fans.

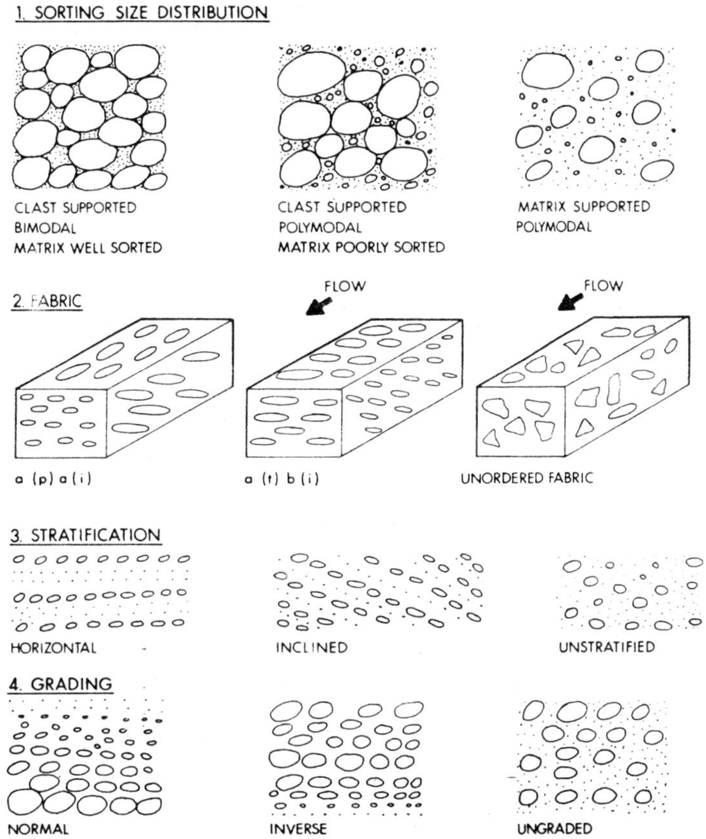

Figure 3.1 Descriptive features for conglomerate. Under fabric, "a" and "b" refer to the long and intermediate axes of clasts, "p" and "t" for orientation of the a-axis parallel to or transverse to flow and "i" for imbrication (from Harms, et al, 1982).

A thorough analysis of flow regime concepts is beyond the scope of this book. A more in-depth, but still succinct treatment of the concepts is found in Harms et al (1982). Suffice it to say, that for a given flow, bedforms which develop through movement of bedload are predominantly a function of flow velocity, flow depth, and mean size of the bedload. Therefore, it is possible to portray the stability fields of various bedforms in three dimensional space (Fig. 3.2). However, such diagrams are cumbersome to read and ordinarily the stability fields are shown on two-dimensional plots where one variable, usually either size or flow depth, is held constant. Possibly the most applicable two dimensional diagram is the one in Fig. 3.3 which is summarized verbally in Table 3.1. In using Fig. 3.3 and Table 3.1 one should not forget that a decrease in water depth can generate about the same increase in scale of bedforms that takes place with an increase in velocity.

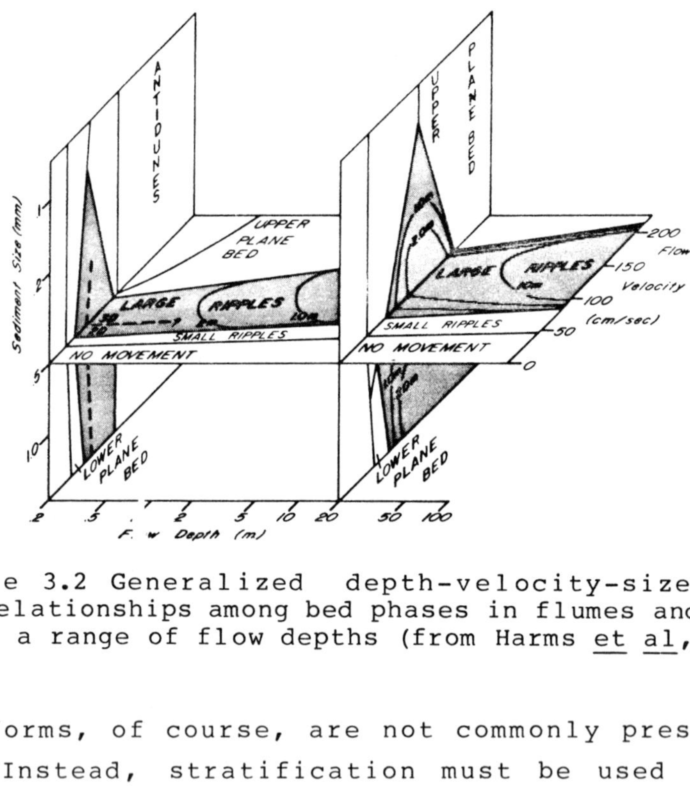

Figure 3.2 Generalized depth-velocity-size diagram showing relationships among bed phases in flumes and natural flows for a range of flow depths (from Harms et al, 1982).

Bedforms, of course, are not commonly preserved in rocks. Instead, stratification must be used to infer bedforms. Typically, small scale cross-stratification signifies small ripples; planar cross-stratification reflects two-dimensional (i.e. straight-crested) large ripples or subaqueous dunes. Trough cross-stratification

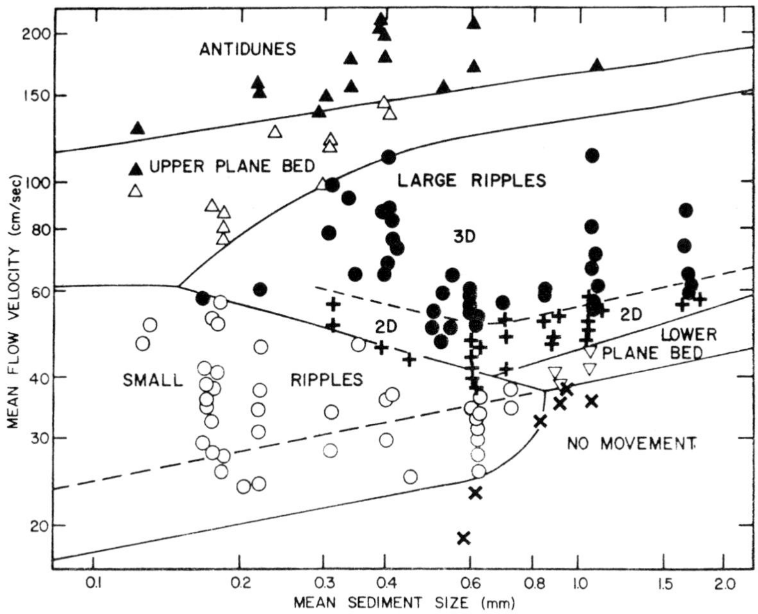

Figure 3.3 Relationship between flow velocity, mean sediment size and regime bedforms (from Harms et al, 1975).

signifies three-dimensional (i.e. sinuous-created) large ripples or subaqueous dunes and parallel stratification indicates plane bed. Antidune stratification is observed very rarely.

FACIES OF CHANNELIZED STREAMFLOW

Introduction

We subdivide the deposits of channelized flow into three facies, (i) channel lag facies, (ii) channel facies proper and (iii) bar facies. Although a wide variety of bar types is recognized in modern braided streams, only a two-fold subdivision into longitudinal and transverse bars seems to have much utility in study of ancient deposits of braided streams.

Table 3.1 Sequence of bed configurations with increasing flow velocity, for various sediment sizes.

SEDIMENT SIZE ↓	FLOW VELOCITY →
silts, very fine sands	no movement — small ripples — upper plane bed - - - - - - - - antidunes
fine and medium sands	no movement — small ripples — 2D large ripples (?) — 3D large ripples — upper plane bed - - - - antidunes
coarse sands and finest gravels (?)	no movement — lower plane bed — 2D large ripples — 3D large ripples — upper plane bed - - - - antidunes

Channelized streamflow deposits on fans are mostly conglomerate, gravel, and granular to coarse sand. Stratification is an almost ubiquitous property. The deposits are long, narrow and oriented quasi-parallel to fan radials. Size grading within individual beds normally is absent (Hein, 1984) and imbrication may or may not be present. If it is present and is not developed within cross-sets, the imbrication is long axis transverse to flow with an upstream dip. Most streamflow deposits are clast supported, but not all clast-supported deposits are the product of stream flow, as we will emphasize later.

Channelized flow on alluvial fans is characterized by declining discharge or waning flood action. This gives rise to the classic fining upward sequences, on a scale of about two meters or less, and the alternating gravelly and sandy strata motif of streamflow deposits (Fig. 3.4). The presence of the fine end member of the sequence is more diagnostic than the absence, in that the upper part of an older sequence may be eroded by a subsequent flow.

Channel Lag Facies

Lag deposits are residual concentrations of pebbles, cobbles, and boulders which have not undergone transport with the sediment which encloses them. The large clasts occur in diffuse sheets which are seldom thicker than the diameter of the largest clast in the sheet. In some cases the diffusion may be so extreme that the lag may consist simply of widely scattered clasts of anomalous size, concentrated along a single horizon and surrounded by waterlaid sediment.

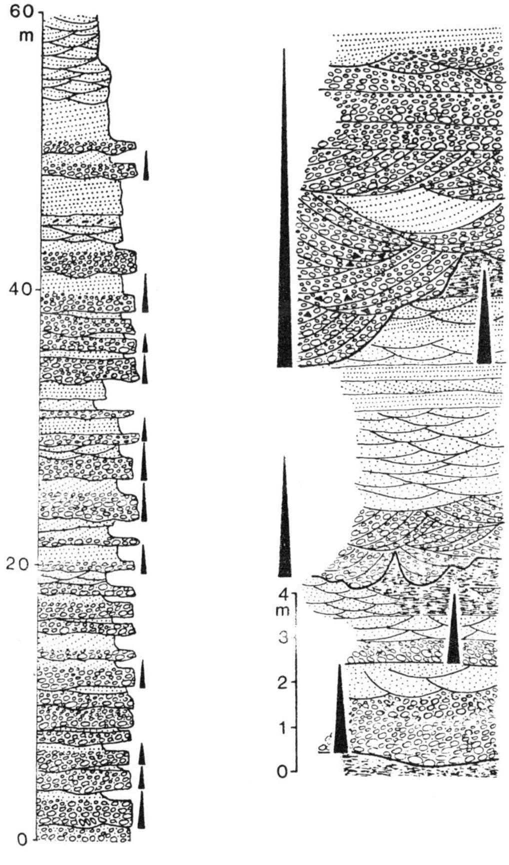

Figure 3.4 Variable scales and facies motifs for fining upward sequences in braided stream conglomerates (from Nemac and Steel, 1984).

Channel Facies Proper

Threads of average largest flow velocity characterize the channels between bars in braided streams. As a result, the dominant bedforms migrating downstream in the channels typically are three-dimensional dunes. These dunes are reflected in sediments by well-developed trough cross-stratification. Depending on the scale of flow, the channel facies will be gravel or sand.

Bar Facies

Longitudinal bar deposits are tabular to lenticular, possibly convex upward accumulations of cobbles and more rarely boulders. Often the conglomerate will be in gradational contact with surrounding water-laid gravel. Stratification is obscure, but, when present, is quasi-horizontal to slightly inclined downstream. Imbrication is common and frequently lamination within internal pods of finer grained sand is faintly visible. Sand wedges often locally mantle the bar conglomerate and possess planar cross-beds dipping at a high angle to the main flow direction. These wedges represent waning flow over the bar and the influence of the bar topography on local flow direction.

Transverse bar deposits mostly form in gravel and sand and are better organized than the longitudinal bar deposits. Well- developed planar cross-stratification, with internal pebble imbrication, is a diagnostic structure of transverse bars. Sets of planar cross-beds may be separated by reactivation surfaces reflecting, in part, erosion during periods of slackwater flow.

Scale

Much is made over the significance of the observed downstream decrease in ratio of longitudinal bars to transverse bars in braided stream systems. Associated with the downstream change is an observed increase in bed relief, mostly because longitudinal bars are lower in relief than transverse bars. Although these trends may be apparent in ancient alluvial plain deposits, it is unlikely that they would exist in alluvial fans. Partly this is due to the shortness in length of braided stream reaches on alluvial fans and partly the flashiness of flow, both of which hinder development of equilibrium flow conditions.

We recommend that more emphasis be placed on the potential significance of trends in thickness of individual cross-stratified conglomerate, gravel or coarse sand beds in alluvial fans. Thickness of cross-sets directly relates to water depth which in turn may be a direct function of channel depth. Cross-sets in gravel and/or conglomerate which are meters thick imply deep channels of the type characteristic of the fan head. These thick cross-sets are produced by unusually large flood events which induce the special process of streamflood (Blissenbach, 1954). Mack and Rasmussen (1984) identify streamflood facies in the Cutler Formation of Colorado. Cross-sets which are centimeters or decimeters thick more likely imply rapidly shifting flow in shallow braid channels of the type more typically found in the mid-fan. Gradients in cross-set thickness, therefore, may be of value in zonation of ancient alluvial fan deposits.

Sheetflood Facies

Sheetflood deposits are laterally continuous sheets of fine to granular sand and gravel characterized by excellent sorting and little clay matrix. They form when flow emerges from the mouths of braid channels, which intersect the fan surface, and spreads laterally or during flood when channels are overtopped and the entire fan is inundated by overland flow (Rachocki, 1981). Because flow is unconfined, it is typically shallow (<0.5 m) implying that plane bed (parallel lamination) and three-dimensional dunes (trough cross-stratification) will be the dominant equilibrium bedforms. Antidunes, reflected by evenly-spaced transverse ribs of coarse gravel may also be common (Wells and Dohrenwend, 1985). Some slackwater deposition of finer grained sand will occur, but these deposits possess a small preservation probability because of erosion by subsequent flood events.

SEDIMENT GRAVITY FLOW DEPOSITS

Introduction

The process which distinguishes the alluvial fan system from all other non-marine depositional systems is subaerial sediment gravity-flow deposition. Furthermore, distinction between humid and arid alluvial fans largely can be based on the presence or absence of gravity flow deposits.

Sediment gravity flows differ from water-induced transport in two ways. First, they possess resistance to shear or strength, which in large part explains their competence to carry large clasts. Secondly, in gravity flows, grains propel the fluid rather than the fluid propelling the grains. Consequently, size and abundance of grains strongly influence flow behavior. A spectrum of sediment gravity flows exists and the terminology and classification can be complex (see for example Lowe, 1979 and 1982). A simple breakdown into two groups, _fluidal flows_ and _mass flows_, is adequate for our discussion which follows (Fig. 3.5).

Fluidal flows are characterized by viscous fluid behavior and low sediment/water ratios (up to 20% water by weight). They deposit by a grain by grain process of accretion on the bed. Fluidal flows include turbidity currents and so-called fluidized flows whose strength is generated by pore-fluid pressures. Mass flows (e.g. debris flows, mud flows, grain flows) are characterized by plastic or pseudoplastic behaviour, and a large sediment/water ratio. They deposit by a process of freeze immobilization. We will emphasize properties and products of mass flows, because they are the dominant gravity-flow process on alluvial fans.

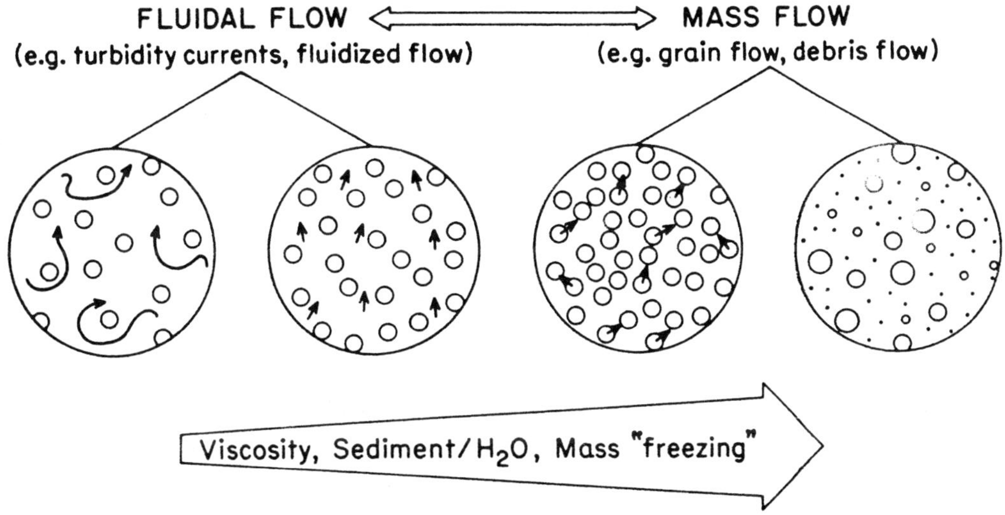

Figure 3.5 Classification of sediment gravity flows. Viscosity, sediment/water ratio and degree of mass-freezing for the various flows increases to the right.

Perhaps the most recent significant development in the sedimentology of mass flows is the recognition of the complex spectrum of flow behaviors which exist, and the increased understanding of the mechanics and variety of deposits associated with the different flow behaviors. Excellent recent discussions of this topic are found in Lowe (1979, 1982), Shultz (1984), Nemac and Steel (1984), and Smith (1985).

Properties of Mass Flows

The standard indicators of mass flow deposition include (a) poor sorting and matrix support, (b) inverse to normal size grading, (c) a-axis clast orientation parallel to flow and imbrication, (d) non-erosive bases, (e) lenticular geometry, (f) absence of stratification, (g) protrusion of clasts into overlying beds and (h) a close correlation between maximum particle size (MPs) and bed thickness (BTh). In reality, all of these properties, with the possible exception of (f) and (h), may or may not exist in mass flow deposits depending on flow behavior. But, observations of MPs and BTh in cores may be of dubious value and the absence of stratification, although virtually ubiquitous in mass flow deposits, also is a common property of many water-laid conglomerates. Further complicating recognition of mass-flow deposits, especially in the subsurface, is the reworking of the deposits by flood processes.

The variety of massflow deposits which exists can be understood in terms of the variable amounts of water, matrix, and framework grains making up mass flows. The variable ratios of those components are a function of source rock, climate, and transport distance and may vary substantially within the same flow, especially as it evolves downslope. Depending on the relative amounts of water, matrix and grains, flows can be characterized by cohesive plastic behavior, viscous fluid behavior, granular collision behavior, or some combination of the three (Fig. 3.7). Flows dominated by cohesive plastic behavior contain a very large mud content, and, therefore, substantial plastic yield strength. As a consequence, their deposits are thick (up to 10 meters), steep sided and narrow (10's of m), and contain very small concentrations of large boulders (Fig. 3.7). Only in the basal zone of maximum shear is any grading likely to be present and it will be inverse. Above the basal zone, material moves as a rigid plug.

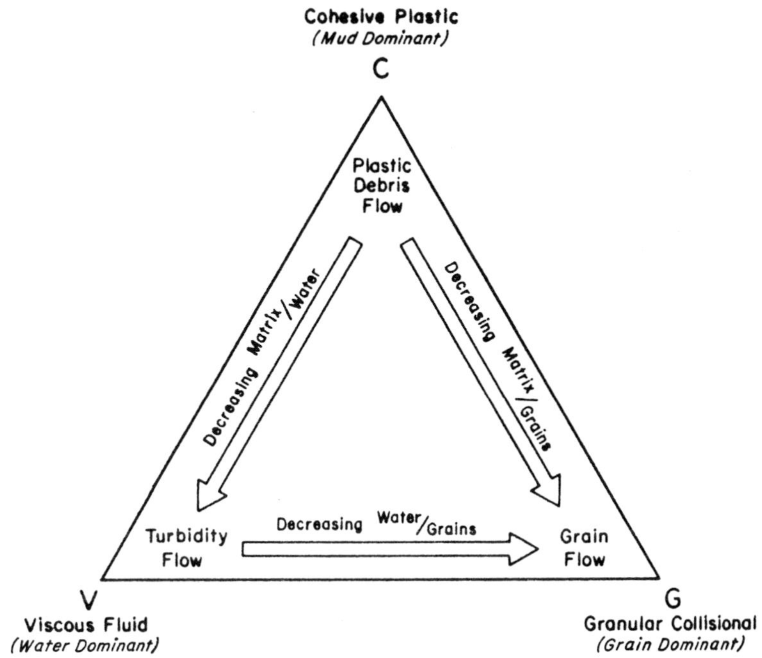

Figure 3.6 Nature of gravity-flow behavior as a function of water, matrix, and grain content of the flow.

Although mass flows are dominated by laminar flow, those containing a large water content locally are turbulent and behave more in a pseudoplastic fashion (i.e. have low yield strengths). The rigid plug layer is thin or absent. Lowered viscosity enhances development of grading, both inverse and normal, the latter developing as the flow decelerates and loses turbulence. The flows lack the strength to carry large clasts and also to maintain thickness. Therefore, their deposits typically are pebbly, rather than bouldery, and sheetlike rather than lenticular.

Flows containing a large volume of coarse clasts dissipate applied shear stress through grain to grain collision which is sometimes referred to as inertial mode of flow (Shultz, 1984 p. 766). Dispersive pressure is exerted on clasts, which is proportionately greater for larger clasts and which is directed away from the bed. As a result, inverse grading of the coarse fraction develops. Deposits are clast supported, unlike classic mass flow deposits, are poorly sorted, and often occupy pre-existent channels, thus their lenticular geometry.

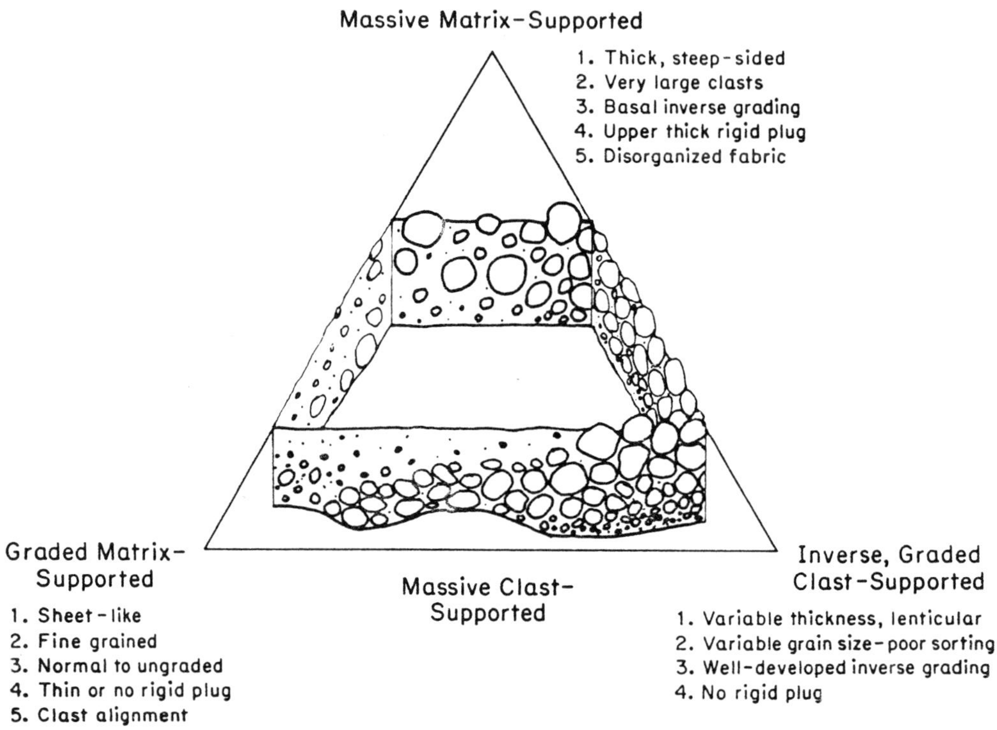

Figure 3.7 End-member types of conglomerates associated with the different types of flow behaviors represented in Figure 3.6 (after Shultz, 1984).

The large matrix content of mass flows precludes any potential as reservoir bodies. In fact, they will behave as permeability barriers. Some general prediction of their distribution and extent in ancient alluvial fans follows from the previous analysis. Massive matrix-supported deposits and inverse-graded and clast-supported deposits should be most common in the upper reaches of the fan, but should be of very limited areal extent because of their lenticular geometry. In contrast, graded matrix-supported deposits, which are a product of more mobile, fluid flows, should be more distal in location, but be thinner and more extensive areally. This areal distribution pattern actually might get enhanced by downfan off-shoots of watery flows from thicker plastic flows on the upper part of the fan as illustrated by Burgisser (1984) (Fig. 3.8). Regardless, systematic downfan decreases in clast size and thickness of mass flow deposits have been documented (Shultz, 1983, Larsen and Steel, 1978). It follows, although not previously documented, that a downfan decrease in lenticularity of mass flow deposits should also exist. These predicted areal trends in types of mass-flow deposits, however, may be greatly complicated in settings where alluvial fans build into standing bodies of water. In these settings resedimentation of coarse clastic material by subaqueous mass-flow gives rise to a totally different population of gravity flow deposits. For more discussion of these kinds of subaqueous flows, the reader is referred to a comprehensive analysis by Nemac and Steel (1984).

SIEVE DEPOSITS

Sieve deposits are a product of a special kind of hybrid flow behavior initially beginning as streamflood and culminating almost as a mass flow. Sieve deposits are standing lobes, up to 10 m thick, of well sorted, angular, open framework debris. They form where little mud is supplied from the source area and accumulate near the mouths of fan-head channels where rapid infiltration of water enters into a gravelly substrate. Coarse material is dropped suddenly forming a barrier behind which finer debris can accumulate (Fig. 3.9). When reactivated during floods, sieve "flows" advance with a caterpillar-type motion through transfer of material as bedload, from upstream to downstream (Ackroyd and Blakely, 1984).

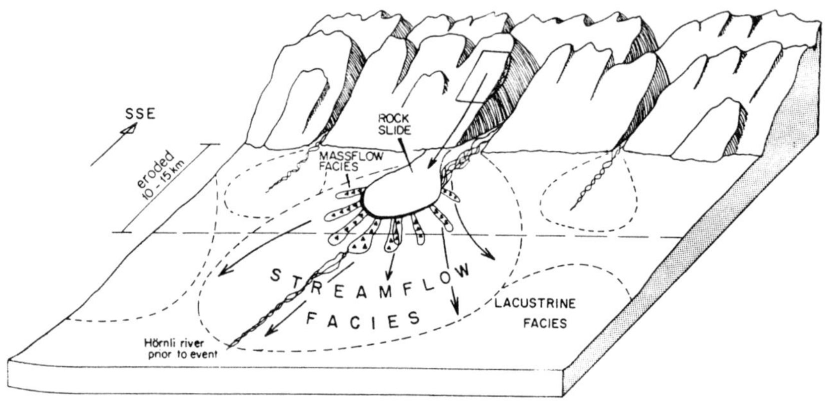

Figure 3.8 Off-shoots of secondary sediment gravity flows from a major gravity slide (from Burgisser, 1984).

Although widely recognized on modern fans, ancient examples of sieve deposits are not common. An Eocene occurrence is reported by Palmer and Merrill (1982) and Gloppen and Steel (1981) describe possible Devonian examples from the Hornelen Basin. Precambrian occurrences were earlier described by Reid (1974) and Pretorius (1979). We suspect that frequent reworking of sieve deposits, coupled with infiltration of finer sediment after deposition, render ancient deposits difficult to identify.

FAN-DELTA COARSE FACIES

Introduction

Proof of the existence of an ancient fan-delta system

demands demonstration of the close lateral and vertical association of subaerial, shoreline and submarine facies. This is easy to do if carbonate, chert, or marine fossiliferous strata occur. It is less easy in their absence. Even if present, the diagnostic facies reveal little of the precise origin of the potential reservoir facies, that is the conglomerates and sandstones. In this section we review some of the diagnostic criterion employed in recognition and understanding the origin of shoreline and subaqueous coarse clastics or diamictite facies in fan deltas. We place special emphasis on those properties useful in core analysis. A comprehensive review of the literature and theory surrounding the topic is beyond the scope of this book.

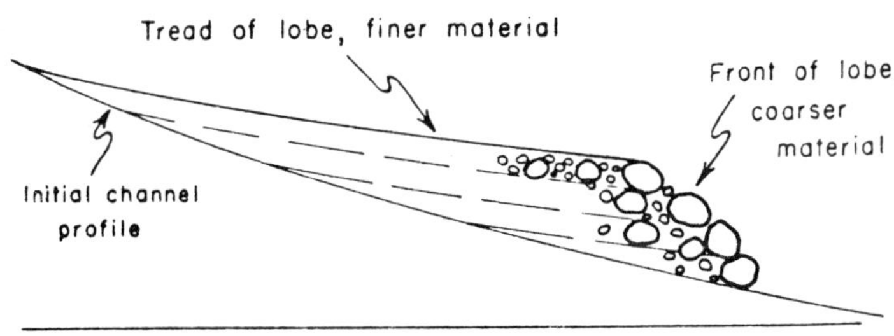

Figure 3.9 Schematic sketch of a sieve lobe (from Hooke, 1967).

Conglomerate Facies Interpretation

Shelf fan-deltas are composed of alluvial and shoreline/shelf facies; slope fan-deltas additionally contain submarine fan facies. This provides a convenient scheme for comparing and contrasting the three genetic varieties of conglomerates that can be deposited in fan deltas, namely alluvial, nearshore, and submarine fan conglomerates. Alluvial conglomerates include both the products of fluvial and subaerial gravity flow; nearshore conglomerates include those produced on beaches or on the shoreface, as well as at distributary mouths, and submarine fan conglomerates include those formed by tractive currents and subaqueous gravity flow. Distinction between the variety of conglomerates often is difficult, if not sometimes impossible. The key observations required for the distinction derive directly from the rocks themselves; therefore, coring and core examination is vital in exploration within ancient fan delta complexes.

Certainly, the first step in understanding the genesis of a complex association of conglomerates involves considering the overall stratigraphic context of the deposits. Next, a careful search must be made for diagnostic associated features, most notably diagnostic facies surrounding the conglomerates. The third step involves careful consideration of the properties of the conglomerates themselves.

Table 3.2 tabulates the diagnostic associated features and facies of the three classes of conglomerates, virtually all of which are applicable to core analysis. Fossil evidence, both trace- and body fossils, of course, is most diagnostic. Next to fossils, associated facies are of greatest value, especially in the case of submarine fans where turbidite sandstones and pelagic deposits should be readily identifiable and abundantly intercalated. The most under-rated of the criteria diagnostic of alluvial deposits are paleosol features. Soil mottling, caliche and micro-soil structures visible in thin-section analysis (Retallack, 1976) should be unequivocal in meaning and abundant. Yet, not a single direct reference to their potential value in distinction between conglomerate associations can be found in the comprehensive volume on conglomerates (Koster and

Table 3.2 Diagnostic associated features and facies useful in distinction between alluvial, nearshore and submarine fan conglomerate sequences.

ALLUVIAL	NEARSHORE	SUBMARINE FAN
1. Paleosols, both arid (caliche) and humid varieties (mottled gleys)	1. Vertical trace fossils (abundant)	1. Horizontal trace fossils (common)
2. Plant fossils and coals (rare to abundant)	2. Eolian dune ss. (rare)	2. Ss with well devel-oped Bouma sequence structures (common)
3. Eolian dune ss. (rare)	3. Nearshore sandstone, lagoonal mudstone and coal (rare to common)	3. Intercalated pelagic and hemipelagic mud-stone and siltstone (abundant)
4. Fluvial ss., slst., and mudstone facies (abundant)	4. Shell fragments (common)	4. Olistostromal beds (rare to common)
5. Small $0^{18}/0^{16}$ ratio in authigenic cement	5. Oscillation ripples mottles, parallel lam-ination, heavy mineral lamination in sandstone (common)	5. Large $0^{18}/0^{16}$ ratios in authigenic cement
6. Mudcracks	6. Glauconite	6. Marine micro-fossils (common)
7. Red beds		7. Gray and dark colored beds

Steele 1984). The most untested, yet perhaps potentially most valuable criterion, may be the oxygen-isotope composition of early authigenic cement in both the conglomerate and sandstone facies. Cement derived from meteoric water should have a lower ratio of O^{18}/O^{16} than cement precipitated from trapped marine porewater and the magnitude of the difference should be directly proportional to distance from the equator. However, sources of complication in this approach are many, and much empirical testing is still required.

Properties of the conglomerates themselves may be revealing, especially if they can be explained in terms of fundamental process differences between the three major sites in which they accumulate. Earlier in this chapter we discussed fundamental differences between water-laid and mass-gravity deposits, and this discussion is applicable to interpreting the origin of conglomerates in fan-deltas. But, fan deltas confront us with unique problems in conglomerate interpretation, namely (i) distinction between water-laid conglomerates of fluvial origin and those of nearshore origin and (ii) distinction between conglomerates of subaerial and subaqueous gravity-flow deposition. We will attempt to make these two distinctions by focusing first on the basic differences in flow behavior and depositional processes in the three sites of conglomerate deposition and then we will link these differences to fundamental differences in conglomerate properties.

River deposition of gravel and cobbles differs from wave- influenced deposition in three important ways: (i) deposition from streams is generally more rapid, with less effective winnowing than takes place along shorelines, (ii) stream flow is channelized in contrast to wave activity which is concentrated over a relatively broad area and (iii) stream flow, particularly braided stream flow, is flashy whereas wave-induced mechanisms of transport and deposition operate more uniformly over larger intervals of time. Rapid deposition by channelized flow gives rise to poorly sorted, organized or segregated beds of conglomerate with lenticular geometry. Waning flow permits infiltration of fine matrix in the upper part of conglomerates, or it may provide a capping of sand to give rise to small-scale fining-upward sequences.

Nearly opposite conditions associated with surf/swash movement of pebbles and cobbles on beaches bring about a winnowing of sand and concentration of clasts in discrete segregated beds, often a single clast size in thickness, of wide lateral extent. However, we emphasize that these relationships are the ideal. Kleinspehn (1984), for example, observed that the degree of organization and segregation of beach gravels can also be indirectly correlated with degree of storm activity which causes disruption on beaches of lag gravels, particularly if a ready supply of coarse clasts is available. Kleinspehn (1984) places more emphasis on the significance of well-developed grading in fluvial conglomerates, but this is not widely supported by our and other's observations. The presence of low- angle stratification is not diagnostic, because this stratification forms on longitudinal bars and on the shoreface.

We tend to downgrade the significance of differences in shape segregation on and between beaches and river channels. Partly this is because shape segregation is difficult to document in cores, but partly it also is because shape is too source-rock controlled. The old adage that shapes are used on beaches, not made there, probably is quite valid. Nevertheless, imbricated tabular clasts dipping in an offshore direction, if apparent, certainly would be diagnostic of beach deposition.

Because of the broad range in gravity flow properties and behavior discussed earlier, and the wide spectrum of deposits which can result, distinction between subaqueous and subaerial conglomerate, on the basis of properties of the rocks themselves, cannot be done with overwhelming confidence. Yet, subtle differences tend to exist and are related to the difference in density and viscosity of subaerial and subaqueous flows. Water, when admixed with gravity flows in a subaqueous setting reduces the density and viscosity of the matrix phase and concentration of the dispersed phase. This accounts for the generally better organization, fabric, and grading of subaqueous conglomerates, as well as their thinner and more widespread distribution. Sudden deposition of a subaqueous gravity flow on a watery substrate also is ideal for the development of load structures, water escape structures and sand injections or dikes (Fig. 3.10). The large water content of subaqueous flows also promotes the formation of turbulent tractive currents of sediment-laden water on top of and in front of the debris flow. These currents, because of their greater velocity, flow in advance of the debris flow and deposit stratified sand which can subsequently be over-riden by the debris flow (Fig. 3.11).

Much has been made in the last five years over the value of plots of maximum particle size (MPs) against bed thickness (BTh) in understanding debris flow deposits in outcrop. BTh/MPs ratios for subaerial deposits tend to be greater than those for subaqueous debris flows (Nemac and Steel, 1984; Gloppen and Steel, 1981) because of the difference in cohesiveness of the flows (Fig. 3.12). Because of the difficulty in defining discrete beds in amalgamated subaqueous debris-flows and in getting a reliable sample of clasts to define MPs, we are pessimistic about application of this technique in core analysis. However, Kessler and Moorhouse (1984) have been successful in applying it to cores of submarine-fan deposits in the Brae field of the British North Sea.

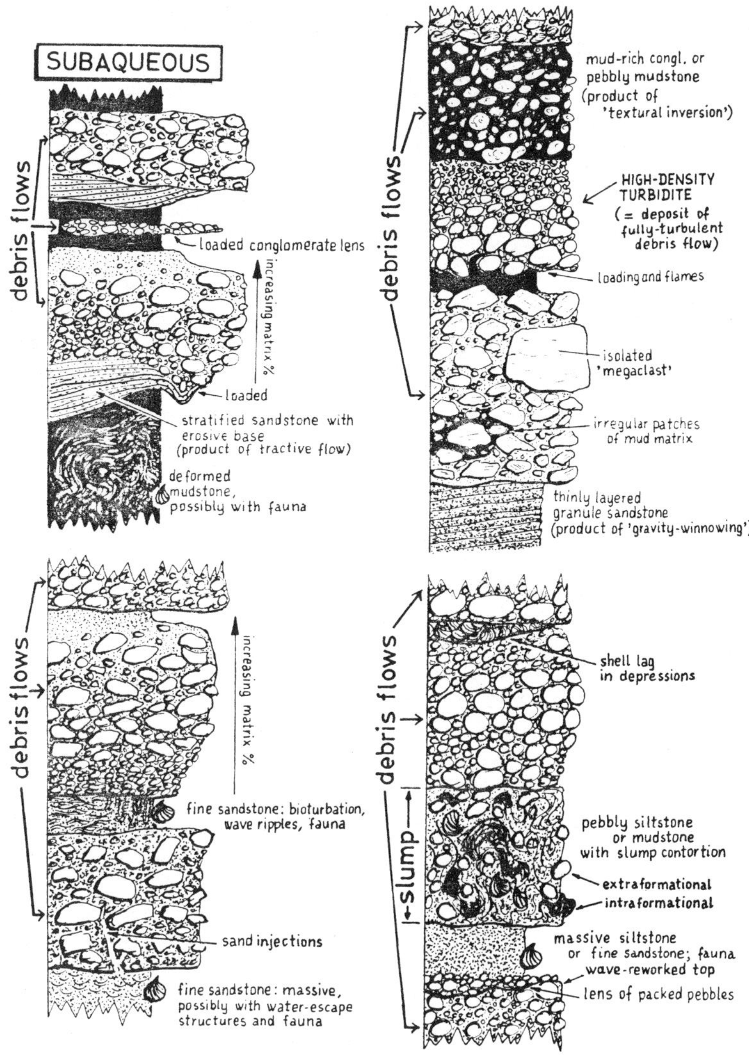

Figure 3.10 Typical features of subaqueous gravity-flow deposits (from Nemac and Steel, 1984).

WIRE-LOG CHARACTERISTICS OF ALLUVIAL FAN AND FAN-DELTA FACIES

One of the frustrating aspects of subsurface analysis of alluvial fan, and, to a lesser extent, fan-delta systems, is the poor wire-log resolution of the different component

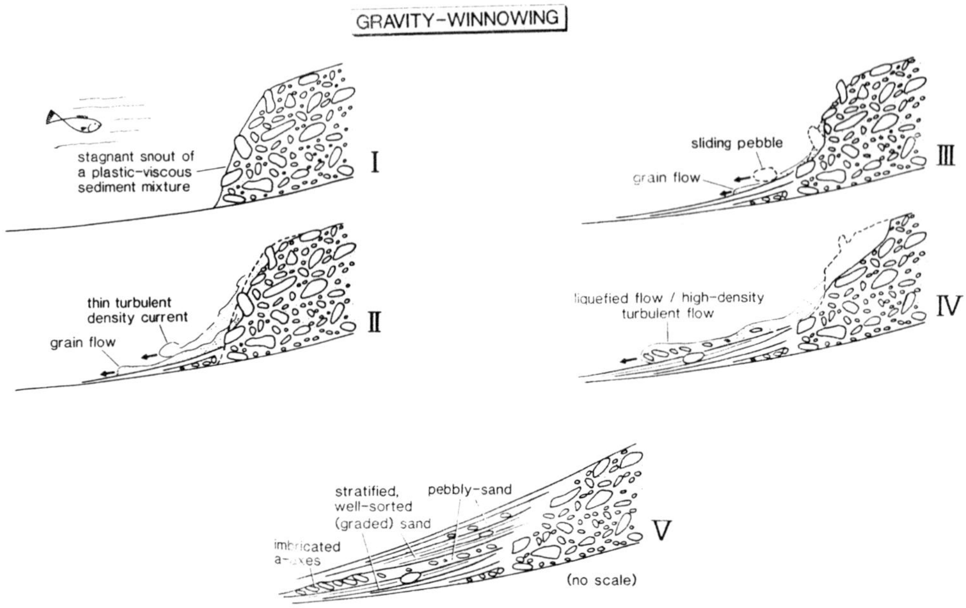

Figure 3.11 Secondary gravity flows derived from a major flow (from Postma, 1984).

facies. This is especially true for the SP and resistivity curves. Partly this is because of the relative lithologic homogenity of the systems (i.e. conglomerate and sandstone often may be the only facies present, and their porosity and permeability will fluctuate widely); partly it also is because there is poor resolution on conventional gamma ray logs of mudrocks owing to the almost ubiquitous occurrence of feldspar and mica in the coarser alluvial sediments. An effective solution to this problem may be spectral gamma-ray logging techniques (Frost, _et al_, 1982). The spectral logs, in addition to measuring total gamma ray emission, measure individual emission from potassium, uranium and thorium. The K, U and Th contents of individual clay minerals, feldspars and micas differ so that differences in spectral log character indeed can be evaluated in terms of clay content of reservoir rocks (Fig.

3.13). However, mud-sized detrital feldspar and mica are present commonly in rocks of immature composition and texture, thereby potentially complicating spectral-log interpretation. Nevertheless, the technique might prove useful in distinguishing (i) mud-rich gravity flow deposits from water-laid coarse clastics, (ii) interfan mud/silt rich facies from coarser fan facies and (iii) levee/overbank facies from trench or channel facies in the fanhead area.

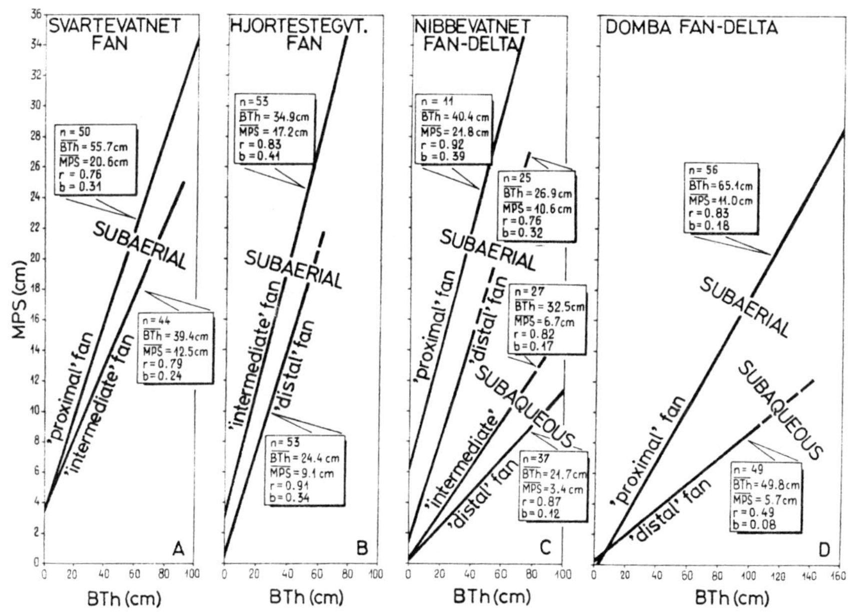

Figure 3.12 Examples of MPs/BTh relationships for subaerial and subaqueous gravity flow deposits (from Nemac and Steel, 1984).

Compensated density-logs also might be useful in distinction between various coarse-grained facies on alluvial fans. For example Candido and Wardlow (1985) demonstrated a positive correlation between amount of detrital garnet plus authigenic pyrite and grain size in a Cretaceous alluvial fan in Brazil. This correlation permitted them to distinguish conglomerate from pebbly sandstone and from finer-grained sandstone on wire logs and was useful in facies zonation of this alluvial fan in the subsurface.

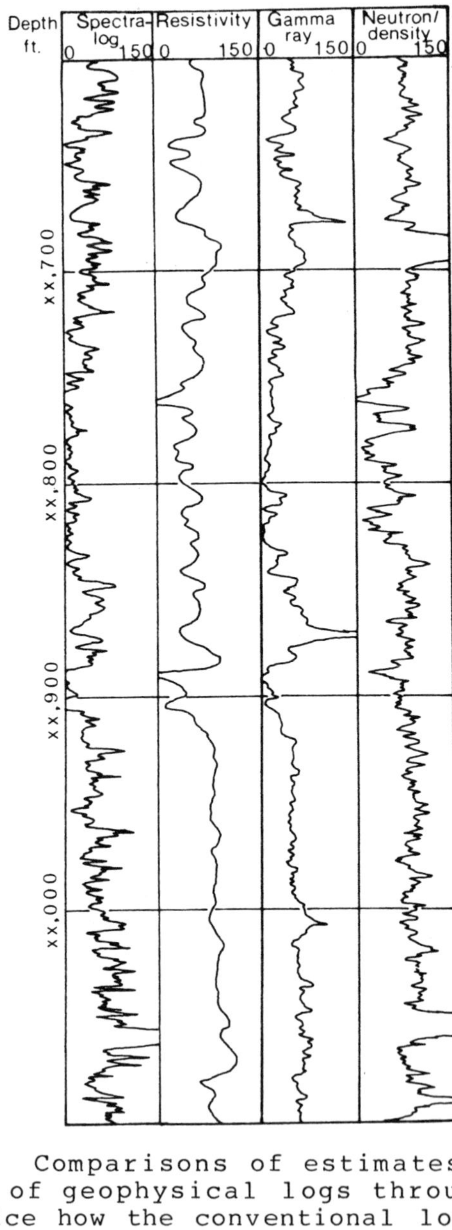

Figure 3.13 Comparisons of estimates of "shaliness" from a variety of geophysical logs through granite wash sequences. Notice how the conventional logs tend to overestimate the amount of shale in the section relative to the spectral log. Shale indicated by dark shading (from Frost et al., 1982).

Better wire-log resolution of facies should characterize the fan fringes and pro-fan environments. Evaporites, lacustrine limestones, and coals or carbonaceous shales found in the continental basin-axis areas should be distinctive. Trunk systems fed by alluvial fans may be characterized by meandering stream deposition giving rise to classical "fining-upward" log profiles.

A significant source of complexity surrounds interpretation of "blocky" E-log signatures for sand/conglomerate bodies in alluvial fan and fan deltas, (Fig. 3.14). This is true even when they cap typical progradational sequences (Fig. 3.15). Avulsion is a common

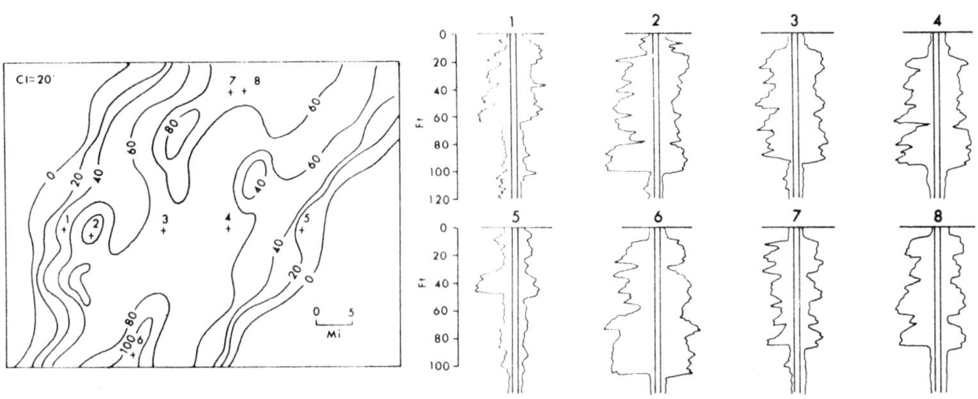

Figure 3.14 Representative electric log profiles of braided-stream channel deposits (from Coleman and Prior, 1984).

process on alluvial fans, and it can quickly cut and strand alluvial and distributary channels, giving rise to sharp upper and lower contacts. Similarly, progradational sequences capped by distributary sands may coarsen and thicken upward and be in sharp upper contact with delta plain or delta front muds, because of avulsion. But, progradational sequences capped by shoreface and beach facies may also be characterized by sharp upper contacts with offshore muds. This is because transgressive episodes are mainly erosional and leave little record, other than perhaps a storm lag gravel at the boundary between the top of the progradational sequence and base of the transgressive "event". This is well illustrated in the fan-delta shoreline facies we have studied in the Fountain Formation

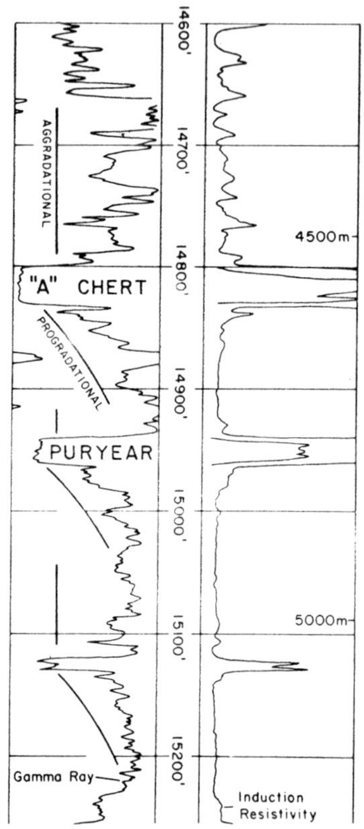

Figure 3.15 Typical coarsening upward, "blocky" electric-log profiles of progradational sequences in Pennsylvanian fan-delta deposits in Texas (from Shelby, 1980).

near Colorado Springs and in progradational sequences described in a Miocene fan delta in California (Clifton, 1981). Therefore, confident distinction between shoreline perpendicular reservoirs (i.e. distributary channels) and shoreline parallel reservoirs (beach/shoreface deposits) may not be possible from electric-log characteristics alone.

4

Geomorphic Controls on Alluvial-Fan Sequences

INTRODUCTION

All thick sections of alluvial fan sediments characteristically show repetitive sequences of various types and scales. These commonly occur in a regular pattern and may consist of cycles that coarsen or fine upward, or thicken or thin upward. Thickness of these cycles may range from less than a meter to 100's of meters, and most commonly, small-scale cycles are superimposed on larger-scale ones.

In general, these cycles previously have been interpreted as responses to regularly recurring variations in extrinsic factors such as climate or tectonism that operate independently of the fan system. We will demonstrate in this section, however, that many repetitively occurring sequences in fan deposits are the result of factors that are controlled by processes intrinsic to the fan itself.

It is particularly important to be able to discriminate between the products of extrinsic and intrinsic factors as they affect fan evolution. The most important sources of external stress on a fan system are climatic change and tectonism, and they tend to produce responses that act similarly over a whole basin. Once the response in one area of a basin is identified, predictions can be made about the response in other parts of the basin. Intrinsic stresses, on the other hand, may trigger responses that do not even affect the entire surface of a single fan. In such a case, complex sediment distributions are produced that can be understood only in terms of integrative models.

INTRINSIC GEOMORPHIC PROCESSES

In order to describe intrinsic geomorphic processes, it is necessary to define several concepts that establish the geomorphic framework in which fans evolve. The first concept defines landforms as dynamic systems that evolve in a way to most efficiently perform the task imposed on them. In this sense, it is convenient to think of an alluvial fan not as a place where sediments accumulate, but as one where sediments are in temporary storage while in transit from source to basin. At any given time, therefore, the shape of a fan is the product of its attempts to maximize sediment transport away from the mountain front.

The second concept is that of the geomorphic threshold. In any system, most physical processes vary in intensity over short intervals, so that system responses are non-uniform and unsteady. Even if processes were to act uniformly and stress was applied to the system continuously, the system might still respond in a non-uniform manner. Quite commonly a system accommodates to stress applied over a length of time until a threshold is reached and a response occurs.

One familiar example of such a threshold is the commonly misapplied term "angle of repose". The angle of repose is actually the slope at which a granular, non-cohesive sediment is stable. This angle is normally several degrees shallower than the internal angle of friction which is the maximum angle that such a sediment can attain before failure occurs spontaneously. The internal angle of friction is a threshold value, and the system is in a metastable state at any angle of intermediate value between the internal angle of friction and the angle of repose.

Response to stress in most geomorphic systems is episodic because thresholds must be exceeded before responses can occur. This implies that systems never attain states of true equilibrium where simple continuity principles apply. Instead, at any given instant in time, a system is evolving toward a threshold or recovering from having exceeded it. Systems that display such behavior are said to be in dynamic equilibrium, and this is the third concept of our framework.

RESPONSE OF FANS TO GEOMORPHIC PROCESSES

An alluvial fan is the product of sediment provided to it by a highland source area, and it in turn acts as a sediment source for the basin into which it is building. As such it is one part of a three-component system. Because these components are mutually interactive, change of state in any one of them can have effects in the other two through feedback processes. In order to understand the controls on facies distributions in fans, therefore, it is necessary to understand the variables controlling sediment delivery to the fan, the processes operating to transport sediment on the fan, and the processes of transport and deposition in the basin.

Source Area Evolution

The major controls on fan evolution are extrinsic

factors, such as climate and tectonism, that affect sediment delivery to the fan. Intrinsic factors within the source area, however, also affect timing and magnitude of sediment delivery to the fan producing responses that are superimposed on the major patterns of sediment distribution established during fan growth.

Following uplift of an essentially smooth surface, drainage systems in a source area evolve in five stages (Glock, 1932; Schumm, 1977) (Fig. 4.1). Development begins with a period of initiation during which a shallow, skeletal drainage pattern is established on the undissected surface. During the period of elongation, enlargement of the pattern occurs along with headward growth of the trunk stream and its principal tributaries. Elaboration of the network occurs as minor tributaries are established in previously undissected areas, and at the stage of maximum extension, the whole of the drainage basin is occupied by stream courses separated by knife-edge interfluves. By progressive reduction of relief, there is a gradual loss of tributaries during the period of abstraction until, eventually, only major drainage lines remain.

From this analysis, we could presume that sediment delivery from a drainage basin might initially be small, increase to a maximum during the period of maximum extension of the drainage net, and then decline through the period of abstraction as drainage density is reduced. The process is intensified by changes in relief that occur concommittantly with source area evolution. Sediment yield increases exponentially as relief in the drainage basin increases (Schumm, 1963; Hadley and Schumm, 1961), probably because the effects of gravity on slope erosion increase with the size of the angle of slope (Schumm, 1977). Relief in the drainage basin is at a maximum and slopes are steepest during the period of maximum extension when the drainage basin is best developed so that sediment production is maximized over the largest available area.

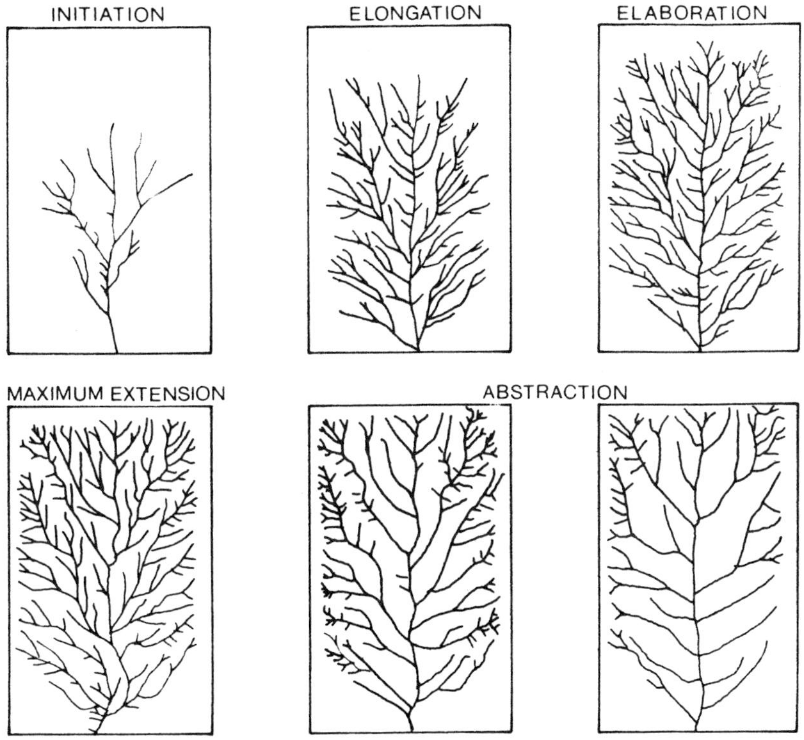

Figure 4.1 Evolution of a drainage basin following uplift of a relatively smooth surface showing initiation (A); elongation (B); elaboration (C); maximum extension (D); and abstraction (E, F); (after Schumm, 1977).

We might expect that the character of the sediment would change as the drainage basin evolved. Sediment size would coarsen rapidly, during evolution, to a maximum at the stage when trunk streams and tributaries were fully developed and slope gradients were steepest. At this stage, streams would be best capable of carrying coarse sediments and the slopes would be capable of providing the coarse sediment. Sediment size would decrease, however, as the drainage basin evolved further. The gravitational component aiding slope erosion would decrease and the degree of chemical weathering would increase as slope gradients declined. Even though their discharge might not vary, the velocity of streams would decrease as their longitudinal profiles declined and the stream's ability to transport coarse sediment would be reduced. More importantly, however, storage capacity would increase as valleys widened, and a special tendency would exist for coarse grades to be stored.

The preceding discussion has demonstrated that it is possible to produce major cycles of sedimentation during fan evolution without recourse to external factors such as climatic variations or tectonism. The sequence that would develop in the absence of these external factors would thicken and then thin upward as well as coarsen and then fine upward (Fig. 4.2). The analysis of drainage basin evolution presented above is based on ergodic studies where space is substituted for time through selection of source areas in different stages of growth (Schumm, 1977). The sequence can also be shown to occur, however, through experimentation (Parker, 1976, 1977) (Fig. 4.3). The experiments show that establishment of the main drainage lines is followed by headward erosion and formation of minor tributaries. Basin evolution reaches the point of maximum drainage density and then begins the period of abstraction as stream density declines.

It is interesting to note, however, that at any given time two stages can occur simultaneously. Maximum drainage density occurs during maximum extension (Fig. 4.3), for instance, but maximum elongation does not occur until the initial stages of abstraction when tributary density on the trunk stream and its main tributaries is beginning to decline as their gradient is reduced.

The tendency for stages of drainage basin evolution to overlap in time has significant implications concerning delivery of sediment to the fan. Even as maximum extension is being achieved at the edges of the drainage basin the trunk stream and its major tributaries are reducing their ability to transport coarse sediment and increasing their ability to store it. Thus, at the same time the basin is producing its coarsest sediment, it is reducing its ability to deliver it to the fan, accelerating the onset of the fining upward part of the cycle and accentuating its effects.

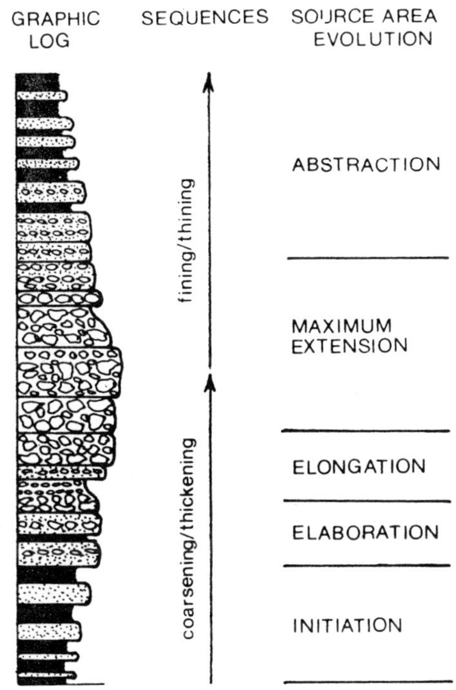

Figure 4.2 Hypothetical sequence that might develop during evolution of a drainage basin.

The onset of a pronounced ability of a drainage basin to store the products of its erosion is actually arrived at relatively early in the drainage basin's evolution, because even in very small drainage basins there is a dramatic downstream increase in the area available to store sediment (Schumm, 1977). With small tributaries of the Cheyenne River basin, for instance, as the drainage area increased only from 0.05 to 2.0 sq. mi., the proportion of valley floor to valley wall increased from 0 to 60 percent (Hadley and Schumm, 1961).

This is also reflected in estimates of sediment yield for basins of various sizes (Boyce, 1975) (Fig. 4.4). In small basins, up to 90 percent of the sediment produced in the basin is removed from it, whereas for basins larger than about 300 sq. mi., only about 20 percent is yielded by the basin and the remainder is stored.

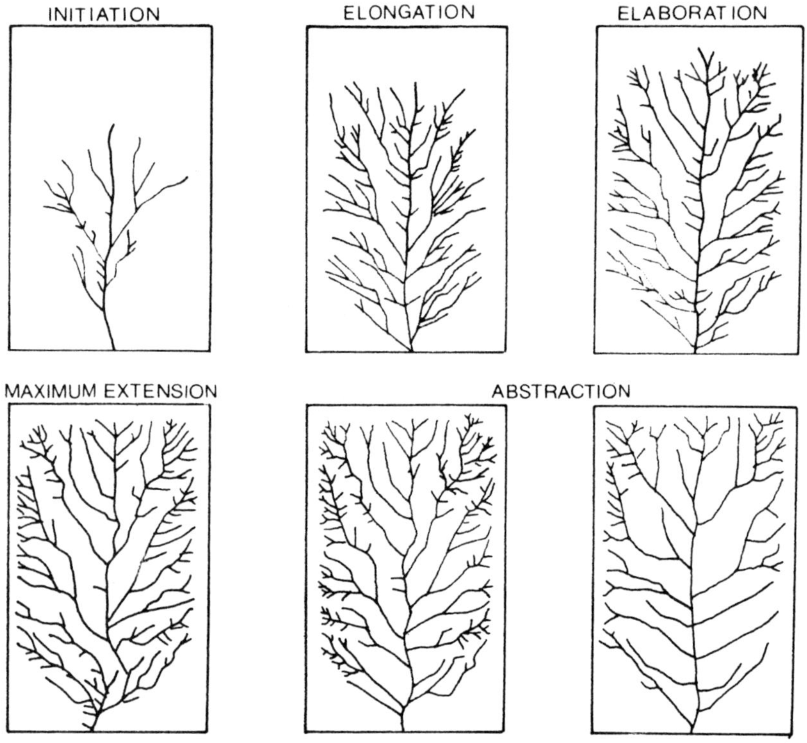

Figure 4.3 Maps of drainage basin stream evolution in an experimental drainage basin (after Parker, 1976).

The pattern of sediment yield from evolving source basins would, of course, determine the quantity and caliber of sediment made available to fans during their growth. Coarse sediments might be delivered to fans during initial stages of basin evolution when stream gradients are steep and the ability of basins to store sediment is limited. As basins develop, however, their ability to store sediment, especially the coarse grades, grows rapidly as valley floors widen and gradients of trunk streams decline. The potential affect on fans is to cause deposition of a fining-upward sequence during the earliest stages of fan evolution (fig. 4.5).

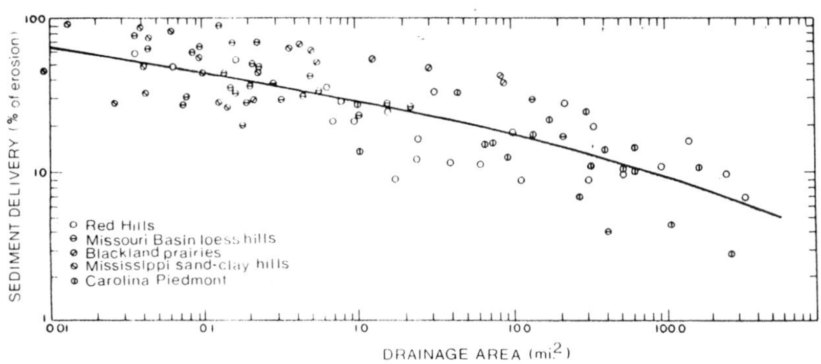

Figure 4.4 Sediment yield as a function of drainage area (from Boyce, 1975).

Continued development increases the total area of the basins, stream channel density, and the proportion of the basins occupied by steep valley walls. The quantity and caliber of sediment delivered to the fans would increase with increasing discharge down the trunk stream and the response on fans might be the accumulation of coarsening and thickening upward sequences during this period. As basins enter their stage of abstraction, however, sediment yield to the fan would decline in quantity and caliber and deposits on the fan might tend to thin and fine upward (fig. 4.5).

Erosion is concentrated at the margins of evolving drainage basins, and even small basins rapidly increase their ability to store sediment. The sediment yield per unit area declines as basin size increases (Strand, 1975)

(Fig. 4.6), but the sediment that is stored is only temporarily retained. Streams may aggrade basin valleys to states of metastability so that changes in extrinsic factors on the fan may trigger an erosional phase when these stored sediments are delivered rapidly to the fan.

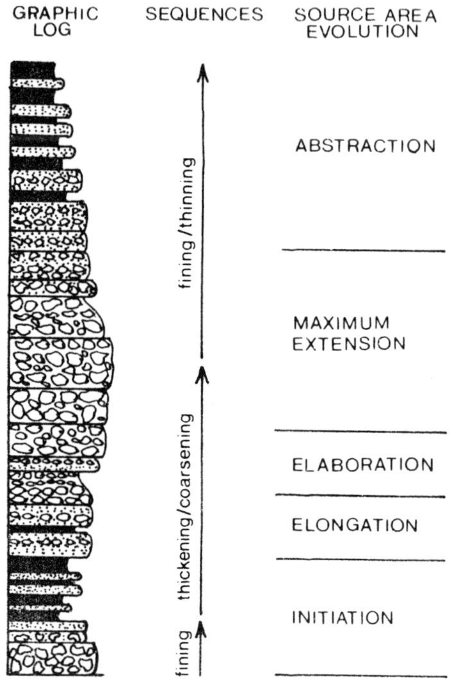

Figure 4.5 Hypothetical sequence that might develop during evolution of a drainage basin. Coarse basal layer is deposited during initial stages of evolution when most sediment eroded from source area is delivered to the fan.

Stored sediments may be eroded if tectonism rejuvenates streams and causes them to incise their valleys. This may similarly occur if climate changes and the consequences of these two intrinsic factors will be explored later. Stored sediments may also be eroded, however, as a consequence of the operation of intrinsic factors on the fan or in the drainage basin itself.

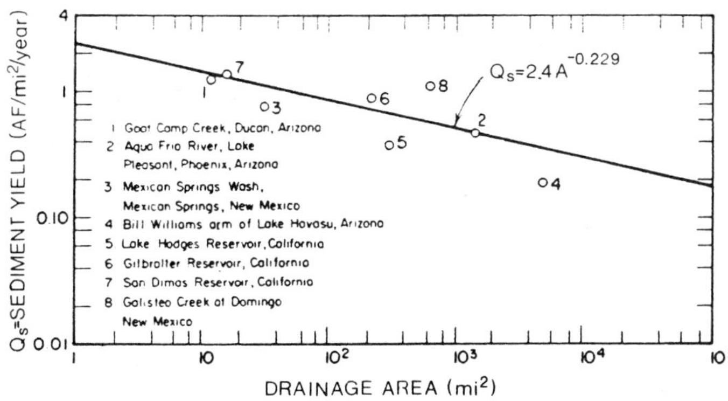

Figure 4.6 Relationship between sediment yield/m^2/year and drainage basin area (after Strand, 1975).

If sediments aggrade valley floors to a point of metastability, a threshold point may be reached when a period of incision is spontaneously initiated. On the other hand, a storm event, that would have had no extreme effects before the valley was substantially aggraded, might trigger such an event under conditions of metastability (Schumm, 1968) (Fig. 4.7). This relationship is shown in an analysis of Piceance Creek, Colorado (Patton and Schumm, 1975). The longitudinal profiles of tributaries to Piceance Creek are

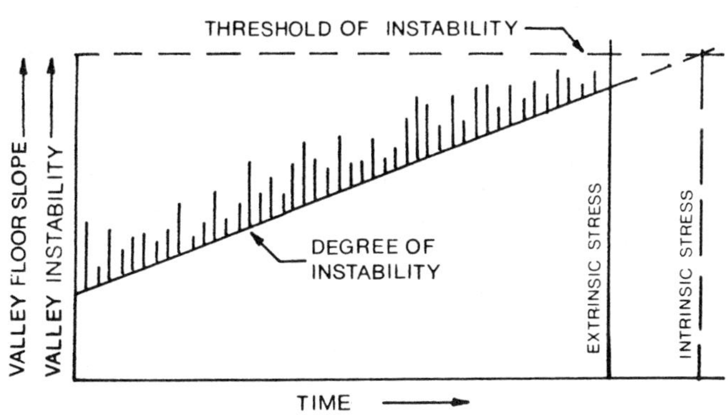

Figure 4.7 Changing valley-floor stability through time showing increasing instability as floor aggrades and approaches the threshold of instability (after Schumm, 1968).

inversely related to the drainage basin size as expected (Fig. 4.8). Two conditions of the valleys are apparent, however, stable ones with smooth surfaces, and unstable ones where gulleys are forming. Unstable valleys have consistently higher valley slopes than those with smooth surfaces suggesting that instability occurred with aggradation. It is also apparent that a gradation does not exist between the two states. Rather, they are sharply defined, suggesting that a threshold condition exists.

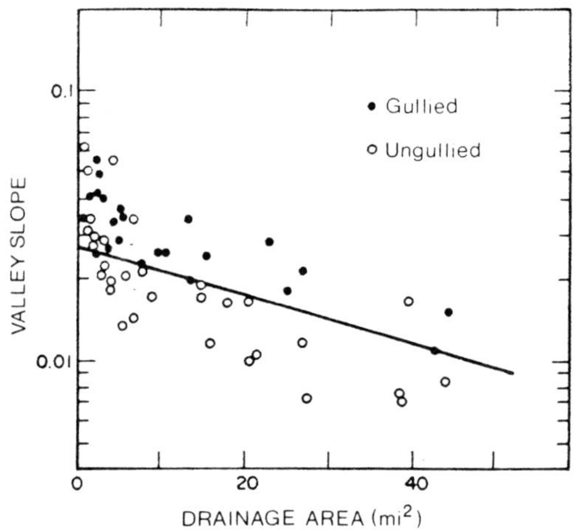

Figure 4.8 Relation between valley floor slope and drainage basins in Piceance Creek, Colorado (after Patton and Schumm, 1975).

The implication that threshold slopes in source area valleys affect fan sedimentation lies in the fact that during the evolution of a drainage basin, valley floors may aggrade to a point of instability, spontaneously incise themselves, and in the process, increase sediment yield to the fan. Not only will the quantity of sediment increase, but its caliber as well, because it is the coarser grades that are preferentially stored in source area valleys.

Thus, geomorphic processes alone acting in the source area can induce cyclic sedimentation of two scales on a fan.

The major cycle is imposed by the establishment and growth of the drainage basin and consists of a coarsening/thickening-upward sequence followed by a fining/thinning-upward one. The minor cycles are superimposed on the major one and are caused by the periodic activation of stored sediments in source area valleys. Initial sediment yields would be large and coarse, but both quantity and caliber would decline with reduction of valley slopes and the resulting sequences would thin and fine upward (Fig. 4.9).

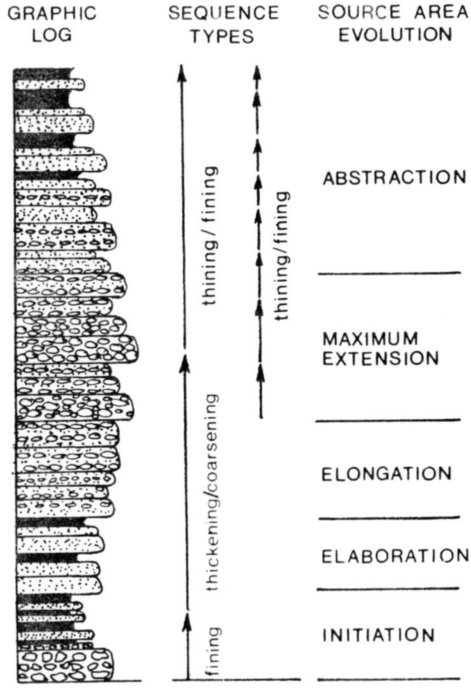

Figure 4.9 Hypothetical vertical fan sequence that might develop through action of threshold mechanisms during evolution of a drainage basin.

Fan Evolution

If we consider fans to be temporary storage sites for

sediments in transit from positions of highest potential energy to positions of lowest potential energy, we can understand the operation of intrinsic factors as they affect fan morphology and sediment distribution. Fans can be thought of as landforms in dynamic equilibrium that are constantly reshaping themselves. Fans evolve in response to changes in discharge and sediment yield from the source area and they reshape themselves to most effectively handle their imposed load. The processes involved in the reshaping include modification of the fan profile by aggradation, trenching, and avulsion of the channel system, and erosion of the fan surface by slope wash and gullying.

One of the primary ways that a fan surface is modified is through fan-head trenching. Fan surfaces vary from those with relatively undissected surfaces where deposition occurs near the mountain front to those where the fan head is trenched and maximum deposition occurs at some point down the fan (Fig. 2.6). Presence of a fan head trench has been considered an indication of change in some external variable such as climate or tectonism, but study of modern fans, as well as experimental studies of laboratory fans have shown that intrinsic stress may trigger fan incision as well.

In Death Valley, for instance, fans vary greatly in morphology and sediment distribution (Hunt and Mabey, 1966). The surface of the Trail Canyon Fan is marked by numerous distributary channels filled with Holocene-age gravels and the youngest of the Pleistocene gravels are evenly distributed, suggesting that deposition is occurring simultaneously over the whole surface (Fig. 2.7). The most recently-deposited gravels on the Hanaupah Canyon fan, however, occur at the fan toe and the youngest of the Pleistocene gravels occur at the termini of several large channels. Deposition is even more restricted on the Johnson Canyon fan where the youngest of the Pleistocene gravels occur in a single trench that crosses the fan from the apex to the foot where maximum deposition occurred.

Such variations in morphology and sediment distribution could conceivably be attributed to differences in tectonic regime from one side of the valley to the other (Denny, 1967). However, tectonism cannot be used to explain differences in fan morphology along the foot of the San Gabriel Mountains in California where fans presumably have developed under similar climatic conditions and tectonic regime (Schumm, 1977). Nevertheless, the Cucomonga Canyon fan has developed a deep trench running the length of the fan, whereas the Deer Canyon fan, adjacent to the east, has no trench.

Eckis (1928) suggested that trenching occurs when sediment yield from a source area declines through time. Incision takes place in order to maintain an equilibrium between transport capability on the fan and sediment influx. If sediment yield declines, but fan gradients remain steep, more sediment will leave the fan than is supplied to it. If trenching occurs, however, the fan profile is reduced through a combination of erosion in the trench and deposition farther down the fan. This limits erosion and promotes deposition.

Hooke (1967) proposed two other mechanisms to account for fan-head trenching. Debris flows dominate fan-head sedimentation in arid regions. They are initially more viscous than water and their viscosity increases by progressive water loss with downslope transport. The debris flows tend to accumulate at the fan head, therefore, forming slopes steeper than those needed to accommodate normal stream flow episodes. When stream flow does occur, therefore, it tends to form channels at fan heads where debris flows dominate.

The second mechanism involves channel avulsion at the fan head following periods of rapid fan aggradation to thresholds of instability. After avulsion occurs, channels shift their courses to follow topographically lower points on the fan. Channels are extended downfan and sediments carried in the channel bypass the fan head and are deposited at the intersection point. Trenches, however, are maintained only long enough to aggrade low areas to a profile in equilibrium with discharge and sediment yield from the source area.

Hooke (1967) demonstrated both processes on an experimental fan that was built both by stream flow and debris flow deposits (Fig. 4.10). Only stream flows were introduced onto the fan during initial stages of construction and the fan surface was aggraded through the period except for six periods when fan head entrenchment occurred as predicted.

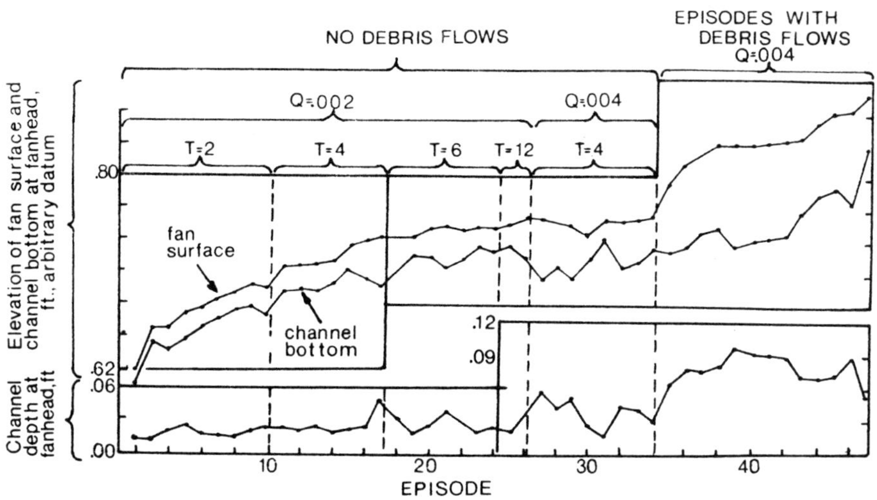

Figure 4.10 Changes in fanhead elevation and depth of incision in a laboratory fan built of both stream flood and debris flow deposits (after Hooke, 1967).

During an intermediate stage, the discharge was doubled. The elevation of the fan-head surface remained constant through several flow episodes, but the fan head was entrenched further. Finally, a series of debris flows were introduced to the fan. The fan head surface aggraded much more rapidly than the fan head channel which became deeply inset within the fan head. At times, the channel became too deeply incised to allow debris flows to overflow channel banks and aggrade the surface at the fan head. The channel began a process of backfilling when this occurred until it was sufficiently shallow for overflow to occur and fan head aggradation to proceed.

These experiments together illustrate a feedback mechanism that establishes a profile and maintains it so that no one segment of a fan can aggrade very far out of equilibrium with other parts of a fan. If incision of the fan head proceeds to a point when overbank deposition can no longer occur, the locus of deposition is shifted downfan and backfilling of the channel above the intersection point begins. The depth of incision decreases until overbank deposition again occurs at the fan head. On the other hand, if overbank deposition at the fan head occurs too frequently because the channel is too shallow, the channel banks aggrade to the point where flood frequency declines and the locus of deposition shifts downfan.

Weaver (1976) and Macke (1976) also designed experiments to simulate evolution of both dry and wet fans. In both sets of experiments, the source material consisted of silty sand mixed with magnetite. The magnetite was used to model the behavior of coarse sediments on natural fans without adding overlarge roughness elements to the shallow flows that occurred on the experimental fans.

To simulate dry fans, mudflows were produced by applying bursts of intense precipitation to the source area, a process not unlike that occurring on natural fans. The resultant fans were conical in shape and showed a regular downfan decrease in sediment size. Magnetite also showed a downfan decrease in occurrence and it was distributed in a regular radial pattern suggesting that little or no reworking of these deposits had occurred (Fig. 4.11), (Weaver, 1976).

As Macke's (1976) wet fan grew, it alternated between two states. In the first, water flowed out over the fan head in a sheet and then was collected into shifting braid channels farther down the fan. In the alternate state, a trench was incised into the fan head, and water supplied to the fan from the source area was conveyed in the trench to distal parts of the fan.

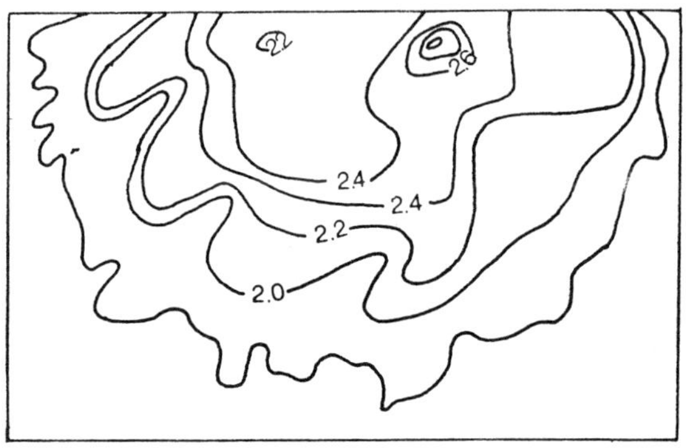

Figure 4.11 Magnetite distribution in a dry experimental fan (after Weaver, 1976).

Apparently, expansion and decelleration of flow occurred when water reached the fan head and spread over it as a sheet flow. Sediments were deposited rapidly at the fan head and the gradient steepened until a threshold was reached and a trench was incised into the fan head. When trenching occurred, the trunk stream in the source area was rejuvenated resulting in an increased sediment yield to the fan which back-filled the trench and restored a smooth configuration to the fan.

Even when discharge was held constant, sediment yield to the fan varied, the fan profile changed from straight to convex, and the fan surface alternated between conditions of aggradation and conditions of erosion and reworking. After the initial surge of sediment from rejuvenated source-area streams passed, backfilling of the trench occurred and the base level of the trunk stream was raised so that valley aggradation was initiated and sediment yield to the fan declined.

The fan-head trenches that periodically developed on the fan greatly affected sediment distribution during the experiments, but other channels formed on the lower fan that also helped to maintain the fan's equilibrium profile and contributed to redistribution of fan sediments. They formed spontaneously on the lower fan during periods when sheet flow occurred, and then worked their way toward the fan head. As they approached it, however, their profiles steepened, discharge and sediment load increased, and they began to backfill and aggrade. The channels rarely affected the fan head although they conveyed large amounts of sediment away from the midfan.

Perhaps the most important results of these studies lie in the way geomorphic processes were shown to affect sediment character and distribution. In contrast to the dry fan, there was an increase in sediment size downfan in response to repeated fan head trenching and to the development of channels lower on the fan. In addition, magnetite, which showed a uniform radial distribution on the dry fan, was locally concentrated laterally on the fan and within certain stratigraphic levels on the wet fan (Fig. 4.12) (Macke, 1976).

Sediment distribution was controlled by the manner of fan aggradation. As water spread over the fan head, deposition was rapid and the fan head grew rapidly and reduced the gradient of the trunk stream in the source area. Storage of coarse sediment and magnetite began in the drainage basin so that fan head growth proceeded with progressively finer-grained sediment. When the fan-head trench was established, however, the sediment stored in the trunk stream was remobilized and delivered to the fan, but because a trench had been established, this sediment bypassed the fan head and was delivered farther down the fan.

The degree to which sediments were reworked during fan evolution also controlled their character. A noticeable contrast developed between sediments deposited during early stages of fan growth and those that were deposited later. Sediments which accumulated early were deposited rapidly and were poorly sorted because they experienced little reworking, and they contained no placer deposits of magnetite. Sediments deposited after the fan head had attained a state of dynamic equilibrium, however, contained miniature placer deposits and they were organized into coarse and fine layers because they underwent repeated episodes of aggradation, and erosion and reworking. In addition, once a condition of dynamic equilibrium was attained, coarse sediment that had been accumulating at the fan head was periodically transfered to positions lower on the fan during periods of fan-head entrenchment and source area rejuvenation.

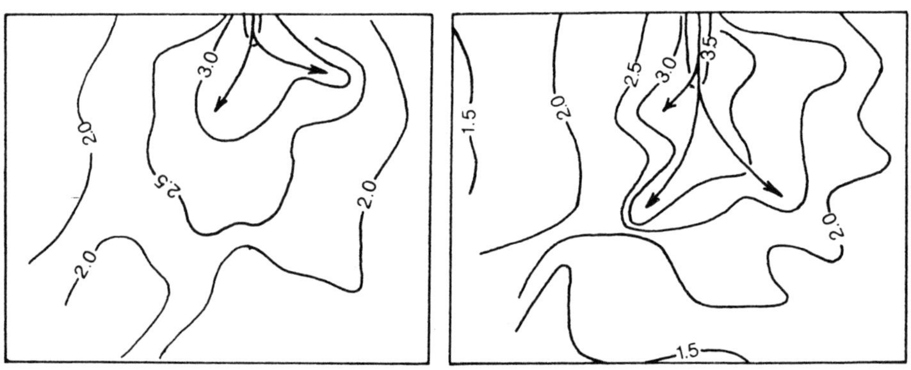

Figure 4.12 Magnetite distribution in a wet experimental fan. Contours are log of magnetite concentration in percent multiplied by 100: (a) 0 to 5 cm depth; (b) 5 to 10 cm depth (after Macke, 1976).

These are rather startling results in that they suggest a degree of self-organization within a fan that most often is attributed to tectonism or climatic variations. This implies that individual fans within a basin will act independently of one another during their evolution. Vertical trends and lateral distributions of sediment types established in one area of a basin may not be applicable to another area where fans have reacted differently to intrinsic stress.

A fan's response to intrinsic stress may have several other consequences that cannot be appropriately modeled in the laboratory. Once a fan has established a profile of dynamic equilibrium it is subject to trenching, and by this process, only small parts of a fan remain active while others become inactive leaving their surfaces exposed to soil-forming processes (Denny, 1965). On dry fans, this most often involves the formation of desert pavements composed of closely-packed rock fragments commonly coated with desert varnish or pitted by solution and perhaps cemented by caliche.

The pavement acts as an armor that promotes runoff and protects the underlying sediments from erosion by water or wind. A silty layer may form immediately below the pavement. The silt may result from weathering in place, or it may infiltrate from above, but in either case, the layer would certainly form a wide-spread impediment to free migration of interstitial fluids (Fig. 4.13).

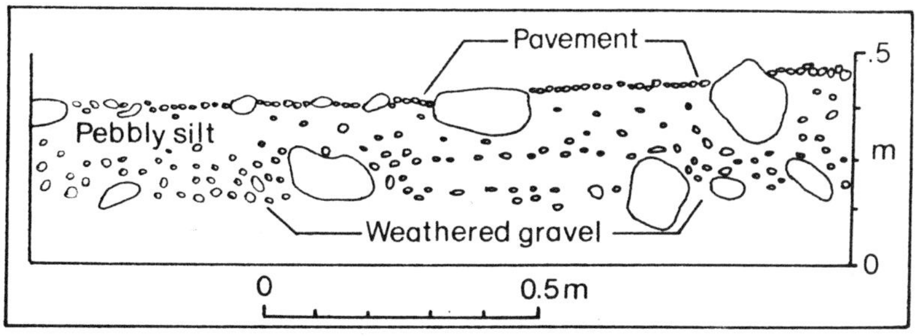

Figure 4.13 Structure and occurrence of pavements on fans in Death Valley, California (after Denny, 1965).

INTRINSIC CONTROLS ON FACIES MIGRATIONS IN FANS

The operation of feedback mechanisms between fan, basin, and source imposes restraints on growth of fans and on lateral migration of facies. The size of both wet and dry fans is closely related to the size of their respective source areas (Fig. 4.14), suggesting that steady-state relationships exist, for a given basin size, between sediment input and removal (Denny, 1965). In cases where either open or closed systems occur, regulating mechanisms exist that prevent the unrestricted growth of fans and the lateral migration and superposition of facies.

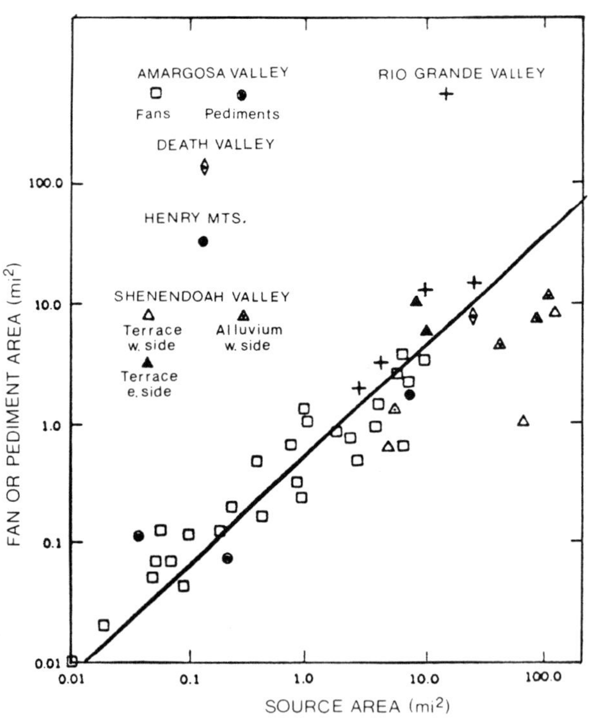

Figure 4.14 Relation between area of a fan and the area of its source area in adjacent mountains (after Denny, 1965).

In <u>open systems</u>, sediments delivered to a basin are

capable of being removed by through-flowing streams. Where this occurs, distal fan facies merge gradually into alluvial plain deposits that are in temporary storage in a system that also is in a state of dynamic equilibrium (Fig. 4.15).

Fan growth is initiated when a source area is elevated relative to an adjacent lowland. Sediment transported from the source area encounters the reduced gradients in the lowland, flow expansion takes place, and deposition occurs. Fan growth continues until it aggrades to the point where the gradients of the fan and the trunk stream of the source area are accordant and a smooth, concave-upward profile extends from the basin floor well up into the adjacent highlands. At this point, the fan is in a state of equilibrium, and as long as the sediment yield to the fan varies only within the limits established by dynamic equilibrium in the source area, the fan will lose as much sediment to the basin as it gains from the source area.

Equilibrium is maintained through several mechanisms. If the fan aggrades too rapidly, a trench is formed and sediment may bypass the upper fan entirely and be delivered to the lower fan and the alluvial plain. Large areas of a fan may also become isolated from a source through avulsion or stream piracy. Once isolation occurs, a process of denudation begins with the formation of small gullies and arroyos fed by sheet wash from the adjacent fan surface. These small channels are capable of transporting only small amounts of sediment, compared to influx from the source area, and in order for equilibrium to be maintained, most of the fan surface at any one time must be rendered inactive (Denny, 1965). The continuing process of channel formation, shifting and abandonment, however, moves the locus of deposition so that fan aggradation occurs evenly over geologically long periods of time.

The deposits that result from these processes consist both of sediments derived from active fan processes alternating with those characteristic of inactive fan

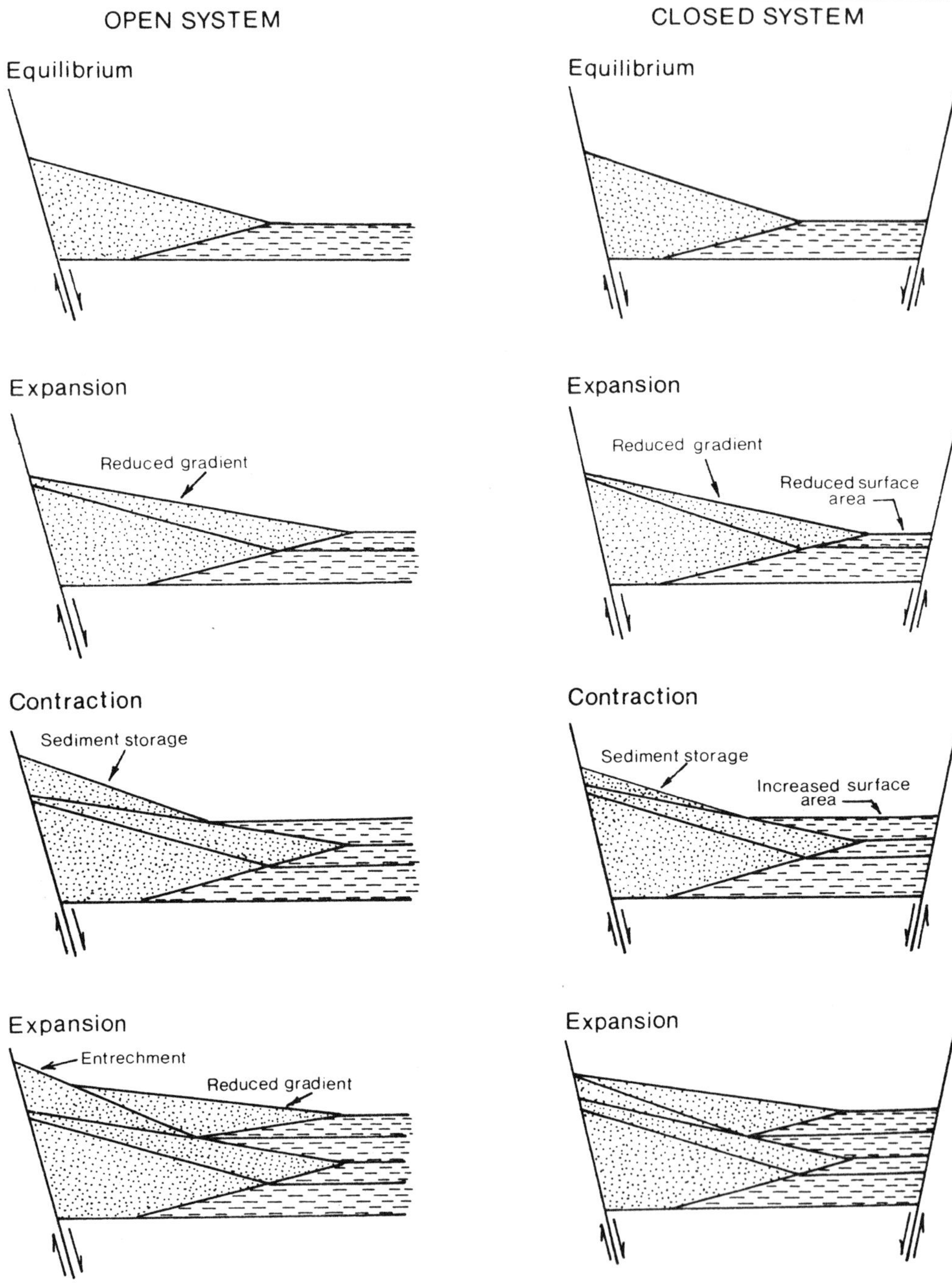

Figure 4.15 Relationship between fan and basin sediments in open and closed basins when both components are in a state of dynamic equilibrium.

processes. Stream deposits occur as lenticular bodies of limited lateral extent along strike enclosed by more extensive, but thinner, deposits of inactive fan surfaces such as lag pavements, caliche horizons, soils, and silt layers.

The equilibrium situation may be disturbed and cause a shift in depositional patterns, but feedback mechanisms will also be triggered so that the disequilibrium is relieved, and the former depositional patterns are reestablished. If sediment yield from the drainage basin increases, for instance, the fan may expand into the basin. In doing so, however, the fan gradient would decline and initiate a period of sediment storage on the fan which, in turn, would cause basin facies to expand onto the fan (Fig. 4.15).

Conversely, if sediment yield to the fan declines, fan head entrenchment occurs, the locus of deposition shifts to a position lower on the fan, and fan facies expand into the basin. The two processes combine to reduce fan gradient and promote retention of sediment on the fan. Streams in the source area are rejuvenated so that additional yield is made available from sediment stored in alluviated valleys in the source area. Again, an equilibrium position is reestablished and extensive retrogradation of the fan is prevented. Indeed, the initial response is a flood of coarse proximal sediments down the fan and an increased yield of coarse sediment stored in source area valleys that combine to initiate a period of fan progration.

Sediment cannot be removed from closed basins because through-flowing drainage is not present. Such basins are commonly occupied by perennial lakes, playas, or eolian plains depending on the local climate. A first approximation would suggest that fan growth and progradation is inevitable in such a situation, but feedback mechanisms also operate in closed basins so that dynamic equilibrium is established after fan growth has been initiated (Hooke, 1968).

If sediment yield to the fan increases, for instance, the fan might extend itself into the basin (Fig. 4.15). This decreases fan gradients, however, and promotes deposition on the fan. At the same time, the area occupied by basinal sediments would contract allowing more rapid aggradation of the basin surface. In effect a buffer is provided against too rapid expansion of fan sediments into the basin.

In a similar manner, the depositional surface of the basin cannot expand greatly at the expense of the fan. As the basin area expands, it aggrades progressively more slowly and, at the same time, reduces fan gradients. Aggradation on the fan accelerates as that in the basin slows so that the equilibrium position is quickly reestablished. In closed basins, therefore, the tendency for fan and basin is to aggrade at nearly the same rate, (Fig. 4.15).

In both open and closed systems, feedback mechanisms impose limits on the extent of lateral facies migrations during fan evolution. Intrinsic thresholds trigger these feedback mechanisms to buffer the effects of changing stress on the system so that conditions of dynamic equilibrium can be maintained. The most important implication of the operation of this principle in controlling fan architecture is that, within the limits imposed by the equilibrium state, facies boundaries will tend to be vertical rather than horizontal (Fig. 4.16). Fan aggradation may not imply fan progradation and the large-scale application of Walther's Law to fan sequences may not be possible.

Response to Short-term Variations in Sediment Yield

Thus far, we have described the response of the fan to relatively long-term variations in sediment yield that result from the establishment of a state of dynamic equilibrium. Fans respond to these variations cyclically and form sequences that measure 10's to perhaps several 100 meters thick.

Fans, however, respond to shorter-term variations in sediment supply caused by storm-induced episodicity of sediment transport and deposition. Even within a steady-state climatic regime, fans are not acted upon continuously. Rather, they experience episodes of sediment movement caused by storm-induced flow variations, and they respond by forming sequences that normally measure less than 1 meter thick. The response on the fan depends very greatly on its climatic setting; storm events produce far different results on wet fans and dry fans.

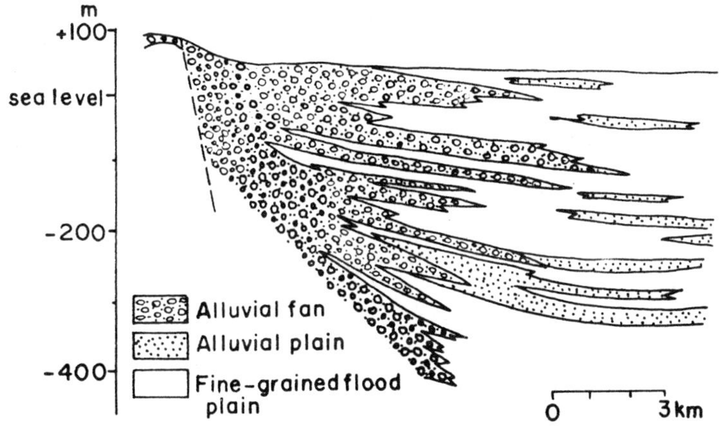

Figure 4.16 Cross-section showing arrangement of Pleistocene-age alluvial fan and alluvial plain facies in the Po Plain of northern Italy. Note the narrow transition zone within which fan and flood plain facies migrate (after Ori, 1982).

Sediments produced by floods on dry fans are highly variable. The erosive capability of flowing water is magnified by the non-cohesive, unvegetated soils of deserts. Under such conditions large amounts of sediment may be incorporated in flood waters and collected in channels forming turbid flows that nearly approximate debris flows. The resulting deposits may be very poorly sorted and matrix-rich with little or no internal organization.

Other deposits are more typical of normal stream sedimentation. Sediments are better sorted and internally well-organized into a variety of sedimentary structures. The bedforms, however, are not well-organized in a predictable manner in response to a hierarchical structure of flow (Karcz, 1972). Rather, they respond to rapidly changing conditions where the flow following flood peak is modified in a complex manner by bed relief.

Erosion of the channel and the formation of lag deposits appears to dominate peak flow, but during waning flow, upper flow regime bedforms are gradually succeeded by lower flow regime bedforms of varying size. Characteristically, larger-scale bedforms act to modify waning flow by imposing some transverse elements on it, and the bedforms themselves are, in turn, reshaped by the flow.

Some floods may exceed bank-full stage producing a sheet flow over the adjacent floodplain. Local irregularities in the surface control sedimentation to a large degree with plane beds forming on highs on the surface, and climbing ripples and tabular cross-bedded microdeltas filling lows (McKee, *et al*, 1967).

Although hierarchical assemblages of bedforms have not been recognized in ephemeral streams, characteristic associations do occur and they result in a variety of vertical sequences. Karcz (1972), for instance, has divided ephemeral stream beds into those characterized by either homogeneous or heterogeneous associations. He found that long reaches of streams were covered by ripples or erosional structures including harrow marks and obstacle marks. Heterogeneous associations include large-scale transverse and lateral bars, both often covered by ripples, large and small scale bedforms modified by erosion during rapid flow changes, and small scale forms deposited by transverse flows that had been redirected by primary bed relief. The deposits of the heterogeneous associations would, therefore, display a high degree of lateral variability.

Picard and High (1973) identified a variety of stratification types that could be grouped into three typical sedimentation units (Fig. 4.17). Point bar sequences are deposited during waning flow and record the resulting progressive decline in velocity and flow depth. These sequences fine upward and consist of low-angle cross-stratification at the base overlain in turn by micro cross-stratification, climbing ripples, and horizontal laminations with a mudfilm at the surface.

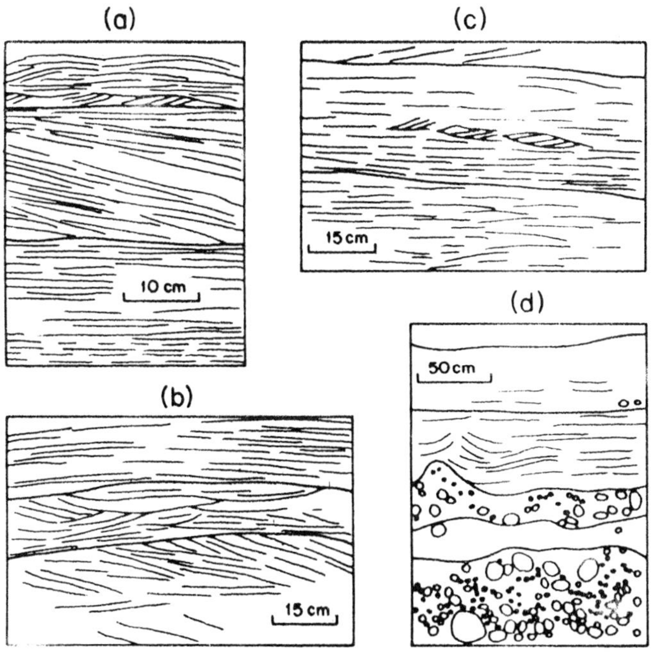

Figure 4.17 Sedimentary structures deposited by flood flow in ephemeral streams (after Picard and High, 1973).

Both longitudinal and transverse bars occur in braided reaches of streams. They manifest themselves in sequences with large scale planar cross-beds overlain by discontinuous plane beds that may, in turn, be overlain by trough cross-stratification, micro cross-stratification, or continuous plane beds. Mud films may also accumulate on bar surfaces where flow has ceased.

Channel deposits accumulate during waning flow by backfilling from distal parts of the channel toward more proximal parts. Basal channel lags may be overlain by upper flow regime plane beds formed just after flood peak. Festoon cross-bedding formed by migrating dunes characteristically makes up the bulk of channel sequences and these may be overlain by micro cross-laminatae or plane beds formed in shallow water near the end of the flow.

Other channel deposits are the result of highly turbid flows that can carry coarse bedloads. Gravelly basal deposits are overlain by poorly sorted fine-grained sediments that accumulate rapidly and maintain high pore pressures so that they are easily deformed.

There are three elements involved in cyclic deposition by ephemeral streams. Individual flood events produce relatively thin packages of sediments bounded at the base by lag deposits and at the top by an upper surface displaying, where preserved, indications of subaerial exposure. Individual sequences fine upward with pronounced coarse and fine-tail grading. Internal structures record the rapid onset of flood, the somewhat less rapid waning, and eventual cessation of flow. In some settings individual storm deposits may be interbedded with eolian sands.

Individual debris flows are also products of storm flow events on arid region alluvial fans. Debris flows or mudflows have long been known to be major components of fan sedimentation. They preferentially occur where source rocks produce sufficient fine materials for their formation. The intense nature of rainfall in desert regions, susceptibility of substrates to erosion, and rapid runoff all contribute to the formation of debris flows.

Flows begin to move only after a critical shear threshold has been exceeded. Movement ceases if shear drops below the threshold due to loss of water, increase in viscosity or decrease in slope. Because of the high viscosity, sediment is carried without sorting and when movement ceases the poorly sorted, often extremely coarse load is frozen in place. Distribution of coarse materials may not be uniform, however, with coarsest clasts often occurring at the front of the flow or on its upper surface. Stream flow commonly follows deposition of debris flows resulting in an intimate association of coarse, poorly sorted debris flow deposits, water-laid sand and gravel (Fig. 4.18).

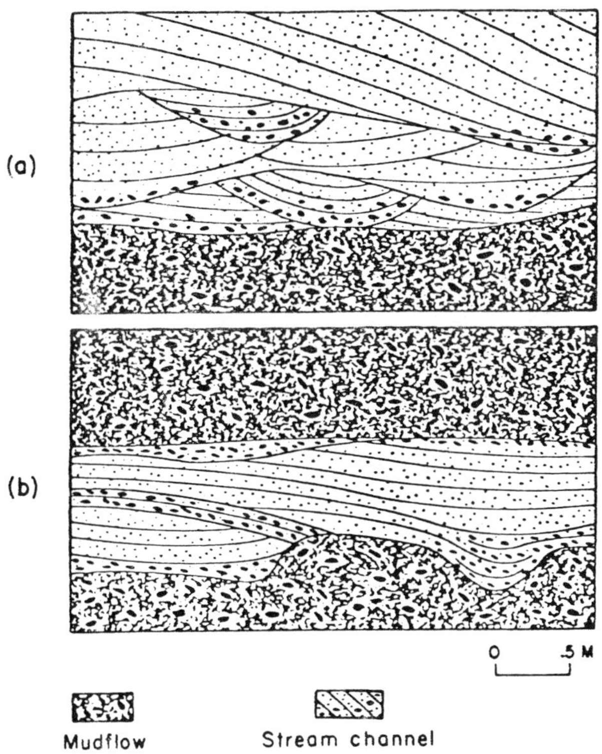

Figure 4.18 Bedding structures in mud flow and stream channel deposits (after Blissenbach, 1954).

Floods frequently percolate into fan surfaces before reaching the toe of the fan. Loads carried by such floods may be rapidly deposited when this occurs leading to formation of lobes of coarse, relatively well-sorted debris termed sieve deposits by Hooke (1967). A sieve deposit may also have a very coarse terminus that acts as a strainer allowing water to percolate through but trapping fines in interstices. Sieve deposits may occur, therefore, where water percolates rapidly into the fan surface or at slope changes where the coarse load of a sheet flood is deposited.

The greatest divergence that occurs between dry and wet fan types lies in the relative importance of channel processes. A single channel usually occurs at the head of arid region fans, but in the midfan area, flow leaves the channel and unconfined flooding occurs. Only during low flow periods is water confined to the shallow channels passing between low amplitude bar forms that were built during flood stage. It is only at the toe of the fan that waters are again collected into significant channels suggesting that braided alluvial channels are distinct from alluvial fan facies.

Channel processes are much more important in humid fans (Boothroyd and Nummedahl, 1978). At the head of a fan several channels may be incised into coarse alluvium. Lower on the fan these channels bifurcate into smaller channels that develop around large bar forms, and farther down fan additional channel splitting takes place around smaller bars. Unlike arid region fans, flow may be maintained in these channels between flood episodes.

Sediment response may be significant but short-lived so that those changes in bar and channel position that can occur on humid fans are not normally preserved. Smaller scale responses include modification of bar shapes, channeling of bar surfaces, infiltration of fine-grained sediment into open framework gravels, and deformation of high angle slopes on bars, all of which occur during waning flow (Fraser and Cobb, 1982).

Storm cycles are relatively short-term phenomena that can have a considerable effect on fan sedimentation. Under such a regime, sediments respond to gradually strengthening flow that peaks at flood strength and then gradually declines. Sediments might be expected to form coarsening-upward and then fining upward sequences as they respond to gradually changing flow, but because of rapid and radical shifts in positions of channels and bars during the cycle, sediment responses may be abrupt rather than gradational.

A typical response to a storm flood cycle might be initiated by the formation and migration of bars or the reactivation of preexisting bars during flood. Localization of flow follows during waning flow and finer grained sediments are deposited in lateral or bar-top channels. The cycle is completed with a second period of bar movement during the next succeeding flood. Fining-upward sequences normally result from such a cycle with coarse bar sediments overlying coarser basal lags and, in turn, being overlain by finer-grained channel deposits (Fig. 4.19).

A second kind of sequence can develop with the reactivation of an abandoned channel which might be related to a flood cycle but which might also be the result of random processes. Where this occurs coarse channel or bar sediments are deposited on top of muds deposited out of suspension during the time flow was not being maintained in the channel. The contact between the two deposits might be sharp or it might be gradational if flow was only gradually reestablished.

Evidence of Intrinsic Controls in Ancient Fan Sequences

Two of the best-documented sequences of arid region alluvial fan deposits occur in the New Red Sandstone (Steel, 1974) and Old Red Sandstone (Bluck, 1967) of northern Britain.

Four distinct conglomerate types occur in the Old Red Sandstone in the Clyde area of Scotland (Fig. 4.20). Mudflow conglomerates consist of poorly sorted, unoriented, pebble to cobble-sized clasts in a sand and mud matrix.

Upper portions of these deposits consist of thin sandstones that probably are the product of reworking of the upper surfaces of flows during the waning stage.

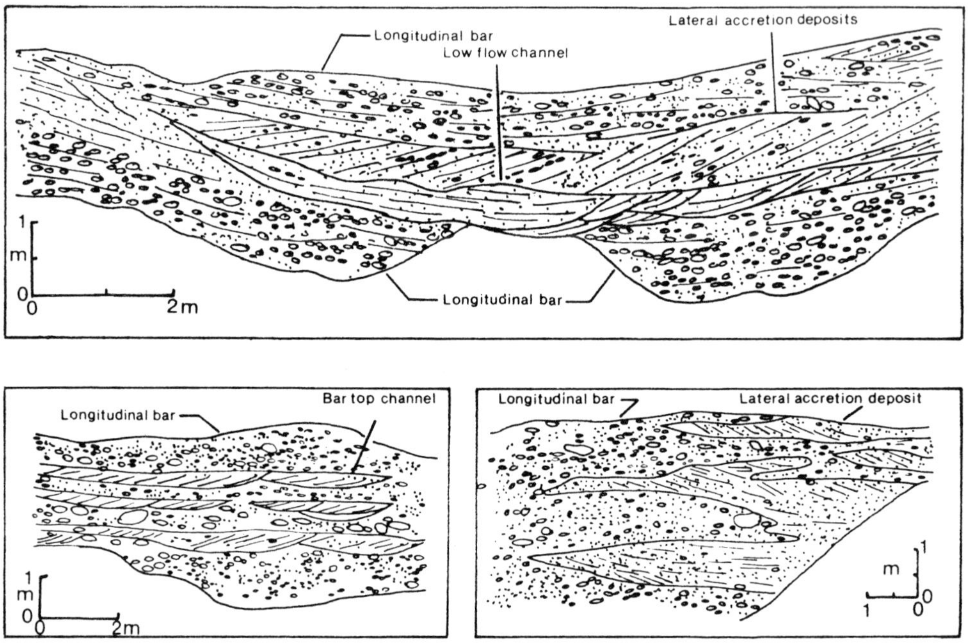

Figure 4.19 Channel deposits accumulating in response to waning flow in glacial outwash fans (after Fraser and Cobb, 1982; Fraser and Fishbaugh, 1985).

Stream flood deposits consist of cross-bedded conglomerate overlain by cross-bedded sandstones occasionally separated by a thin lag deposit. Basal contacts of the conglomerates are erosional. The coarse-grained basal unit probably accumulated during maximum flow and the cross-bedded sandstones represent late-stage reworking during waning flow. Typically these deposits are associated with mudflow conglomerates.

Braided stream conglomerates consist of cross-bedded sand and gravel and occasional beds of plane bedded sandstone and intraformational conglomerates. These beds appear to have formed in a setting characterized by rapid lateral channel shifts and their local association with stream deposits suggests a depositional process not unlike sheet floods. Finer grained units are commonly associated with playa deposits suggesting a distal fan or basinal setting. Other basinal deposits include floodplain conglomerates that occur as thin lenticular units enclosed in sandstone. They probably represent lag deposits in channels on the floodplain.

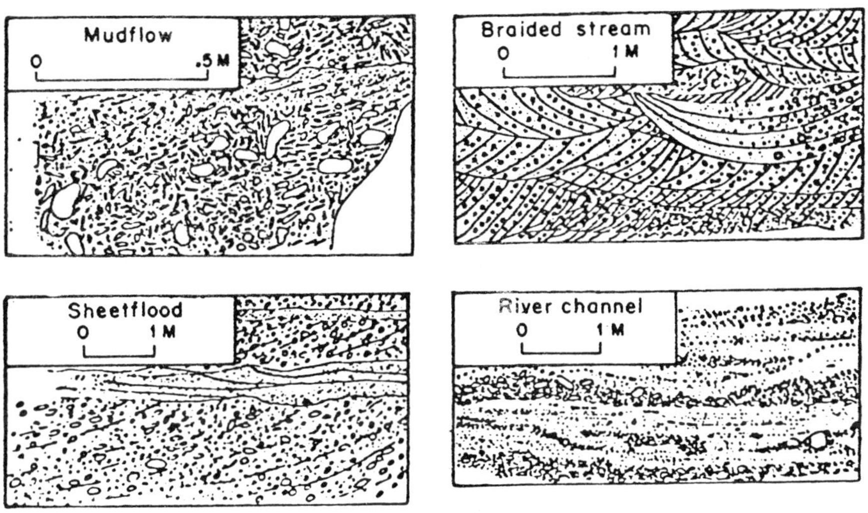

Figure 4.20 Conglomerate types in the Old Red Sandstone of Scotland (after Bluck, 1967).

The vertical sequence shows a pronounced fining upward trend (Fig. 4.21), and there is also a change upward from mudflow conglomerates interbedded with playa mudstones, to braided stream and stream flow deposits. Bluck (1967) interpreted this sequence to represent the diminishing effect of a receeding source. In this interpretation, fan deposition is initiated with proximal mudflows accumulating in close juxtaposition with playa deposits. As the source recedes, both are overstepped by braided stream conglomerates and floodplain deposits.

The presence of basinal deposits well up in the

sequence is not easily explained with this hypothesis but an alternative hypothesis can be offered, however, that explains the sequence in terms of source area evolution. During initial stages of fan formation, the drainage basin was small, valley slopes were steep, and there was little room in valley bottoms to store sediment. Most of the sediment produced in the basin was yielded to the fan, and because the gradient of the trunk stream was steep, it was capable of transporting relatively coarse material. These early stages of drainage basin evolution may be reflected in the base of Bluck's (1967) sequence by the presence of coarse-grained mudflow deposits indicative of proximal processes (Fig. 4.21).

With continued basin development, its storage capacity increased and the gradient of the trunk stream declined. Because of this, the rate of sediment production as a function of drainage basin area declined and coarse sediment was preferentially stored in source area valleys. The response on the fan was the accumulation of mostly finer-grained sediments and the source-ward migration of basin facies. This stage of basin evolution may be represented in Bluck's (1967) sequence where finer-grained conglomerates are interbedded with basinal marls (Fig. 4.21).

As the drainage basin reached maximum extension, sediment yield to the fan increased, but coarse sediment continued to be trapped in expanding valley floors. The result was an increased yield of relatively fine-grained sediments to the fan and the basinward migration of fan facies. In addition, as the basin evolved toward a larger and more fully integrated drainage network, discharge became more regular and channel processes began to dominate fan processes. This stage is represented in the upper part of Bluck's (1967) sequence by 60 meters of braided stream sediments that become progressively finer upward.

Similar sediments occur in the New Red Sandstone of the Hebrides (Steel, 1974). Conglomerates with a silt and

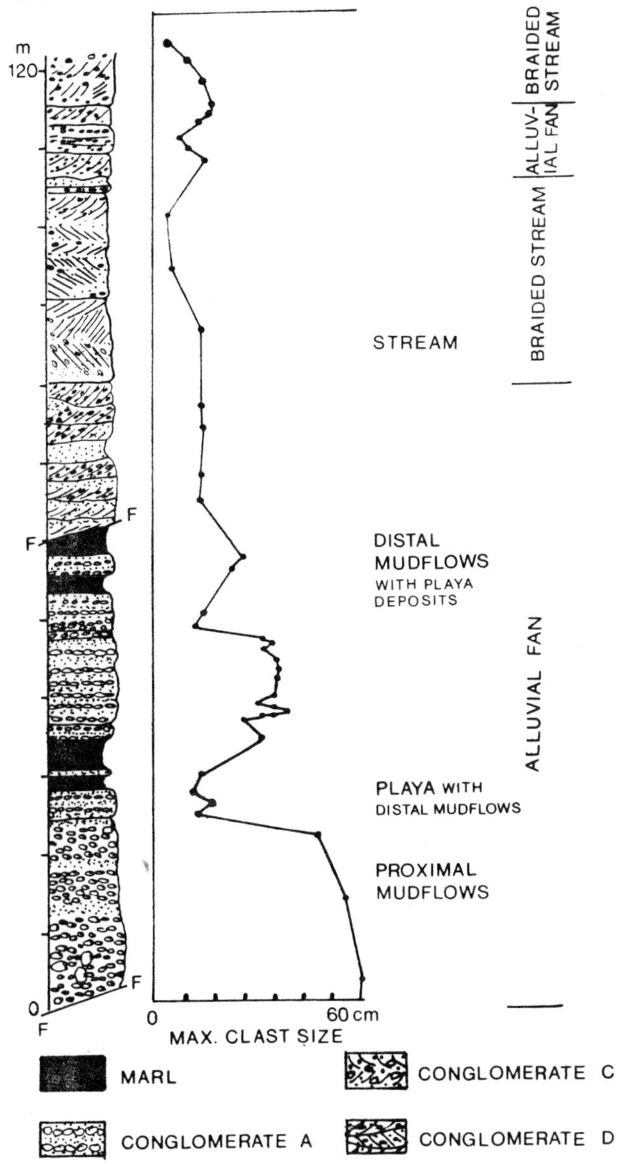

Figure 4.21 Stratigraphic section through the Old Red Sandstone at Portencross, Scotland showing overall fining-upward sequence (after Bluck, 1967).

clay matrix are interpreted as mudflow deposits (Fig. 4.22). Discontinuous beds of well-sorted, clean conglomerate, often in large scale crossbeds, are normally associated with the mudflows and apparently formed as flood stage channel deposits. Braided stream deposits are well-sorted but they are characterized by rapid lateral variations in texture. They are finer-grained than stream flood conglomerates and may have accumulated in channels heading on the fan surface forming lenticular deposits (Fig. 4.22). Other similar deposits are more laterally continuous and may represent sheet flood deposits of depositional lobes.

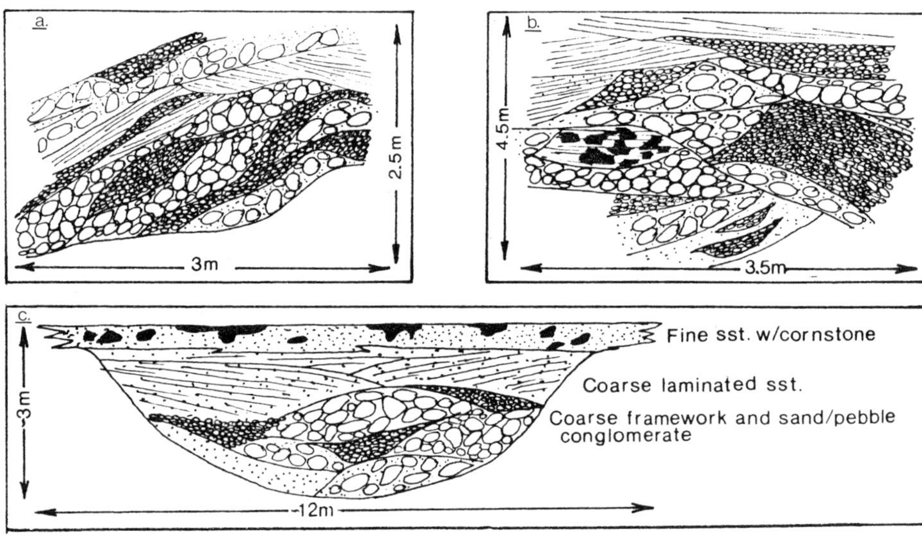

Figure 4.22 Braided stream sediments (upper) and a channel-fill sequence (lower) in the New Red Sandstone of the Hebrides (after Steel, 1974).

Subordinate facies include playa deposits commonly associated with distal components of mudflows and stream floods, and caliche crusts found in sandstones capping braided stream deposits. The occurrence of both facies is suggestive of an arid climate.

In addition to alluvial fan sediments, floodplain deposits of both high and low-sinuousity streams also occur in the New Red Sandstone. They may occur at the base of some sequences where they are in sharp contact with overlying fining-upward alluvial fan sequences, but in other sequences, alluvial fan deposits are capped by floodplain sediments (Fig. 4.23).

Individual alluvial fan sequences commonly show a preferred sequence of facies. They may consist of mudflow deposits overlain by braided stream or stream flood deposits (Fig. 4.23a). This upward change in sediment type is accompanied by an upward decrease in clast size, an increase in percentage of sandstone beds, and an upward decrease in the ratio of extraformational to intraformational clasts.

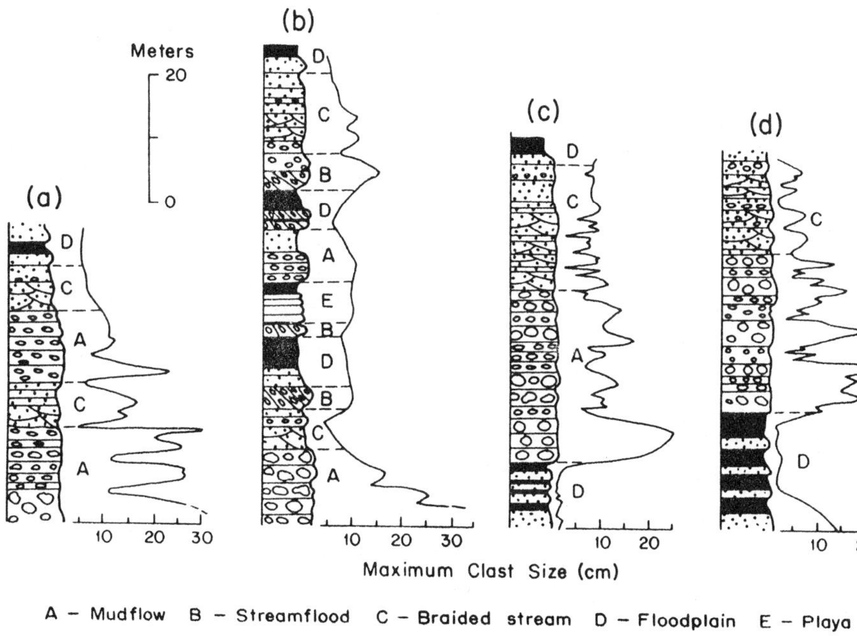

Figure 4.23 Variations in vertical sequences in the New Red Sandstone of the Hebrides (after Steel, 1974).

Other sequences are similar to those of the Old Red Sandstone described by Bluck (1967). They consist of coarse-grained proximal alluvial fan sediments at the base, a middle part of fine-grained playa and floodplain sediments, and a fining-upward sequence of alluvial fan sediments at the top (Fig. 4.23b). These sequences probably formed along basin margins and represent the complete sequence of source area evolution.

The New Red Sandstone in other areas consists of fining-upward alluvial fan sequences over thinner deposits of floodplain or playa sediments (Fig. 4.23c, d). The basal fine-grained portion of these sequences probably was deposited during initial stages of fan formation when basin facies were deposited near the basin margin. Basinal facies coarsen upward through a short transitional interval into a fining-upward sequence of alluvial fan sediments that were probably deposited during the remaining stages of drainage basin evolution.

Steel (1974) also interprets such sequences as possibly having formed in response to a recession of the source area, but he also attributes their origin to long-term changes in the source area including a lowering of relief. As in the case of the Old Red Sandstone, storage of coarse sediments in expanding source-area valleys very likely also played a part in origin of the fining-upward sequences.

Steel (1974) also recognizes significance in the change in clast composition upward in each sequence. He suggests that the increase in percentage of intraformational clasts signals the progressively growing importance of channels originating on the fan surface as envisioned by Denny (1965).

De Feyter and Molenaar (1984) differentiate six lithofacies (Fig. 4.24), in fanglomerates of Messinian age in Italy, that are arranged in regular sequence both laterally and vertically. They attain a total thickness of 850 m and occur in eight wedge-shaped bodies of conglomeratic sediment embedded in marls. Unlike most other fan sequences, these had their source in a fold belt rather than a fault block.

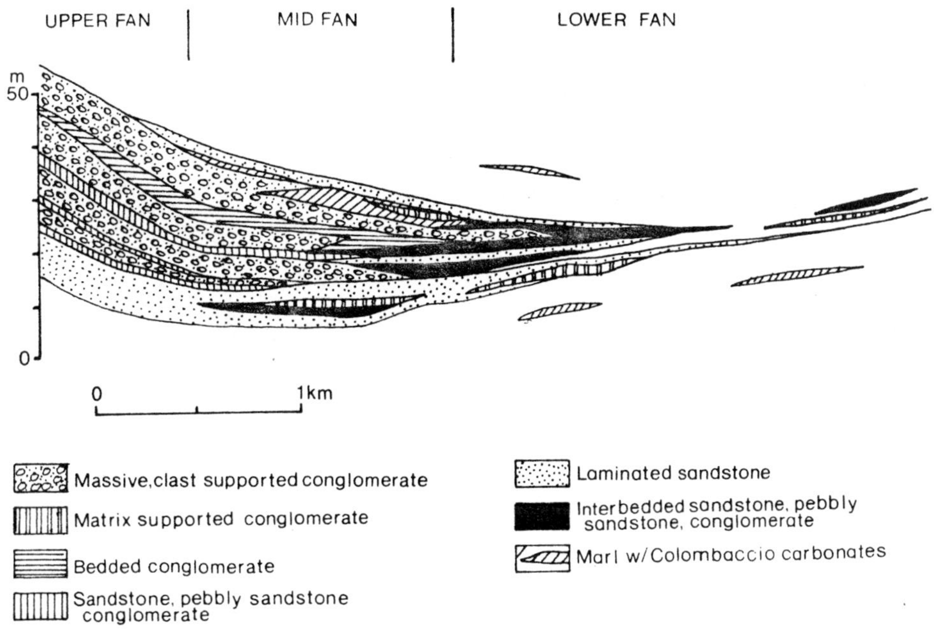

Figure 4.24 Synthetic section through an alluvial fan body based on six measured sections (after de Feyter and Molenaar, 1984).

Deposits nearest the fold belt are dominated by debris flows that consist both of massive, disorganized conglomerates that are clast-supported in a sandstone/mudstone matrix, and matrix-supported conglomerates in inversely-graded beds. Interbedded with debris-flow conglomerates are bedded conglomerates that were probably deposited on longitudinal bars. They may record intermittent stream-flow episodes, or partial reworking of debris flow sediments.

Bedded conglomerates grade laterally basinward into cross-bedded sandstones, pebbly sandstones, and conglomerates deposited as transverse bars, dunes, and sandwaves in a braided stream. They occur in lenticular beds bounded by channel scours and may represent channelized flow characteristic of mid-fan processes.

Thin-bedded, laminated sandstones occur at the base of

sequences and also as interbeds with other facies distal from the source. They occur as sheetlike layers that display horizontal laminae or shallow scour-and-fill crossbedding characteristic of unconfined sheet-flow processes. They probably represent deposition where rapid flow expansion occurred such as at the fan toe or on the basin plain.

De Feyter and Molenaar (1984) interpret the lateral and vertical distribution of facies to be the result of fan progradation, yet the columns and synthetic cross-section they display do not suggest simple fan progradation (Figs. 4.24 and 4.25). The intermediate-fan section, for instance, is bounded both at the bottom and top by deposits more characteristic of distal fan or basin plain processes. The conglomeratic facies that do occur are concentrated in the lower 2/3 of the sequence.

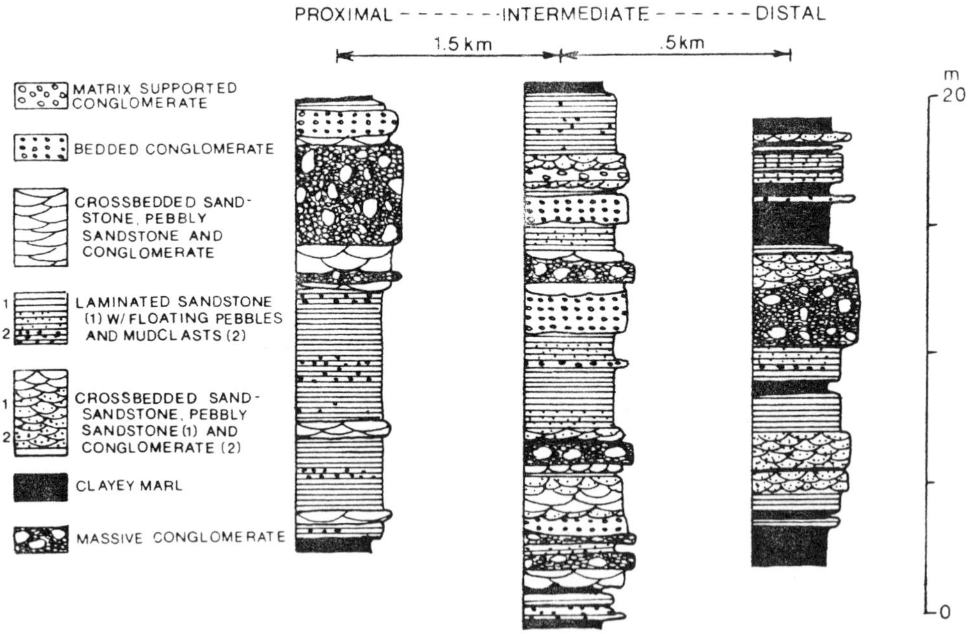

Figure 4.25 Vertical sections characteristic of different reaches, of Messinian fanglomerates (after de Feyter) and Molenaar, 1984).

The proximal sequence does contain thick debris flow deposits, but these occur only at the top of the sequence and the bulk of the section consists of thin-bedded sandstones of the fan toe/basin plain facies. The distal section is dominated by thin-bedded sandstones and marls, as expected, but here they overlie and underlie debris flow conglomerates that are much thicker than any occurring in the midfan section.

It is also apparent from the cross-section that these fan bodies were relatively small; true fan deposits did not extend more than 4 km from the source, and proximal fan deposits did not extend, for the most part, beyond 2 km from the source. Rather than a progradational or overlapping pattern, it is easier to draw a vertical cutoff between proximal and mid-and distal fan facies. In addition, a retrogradational pattern is evident at the top (Fig. 4.24).

These features can more easily be interpreted in terms of the controls that source area evolution exerts on fan growth. Initial deposits of the sequence consist of laminated sandstones with outsized clasts. These probably were deposited during the first stages of source area evolution when relatively fine-grained sediments were being delivered to a basin whose floor occurred very near the source area.

Stages of elaboration and maximum extension of the source area are represented by deposition of coarse debris flow conglomerates on the upper fan, finer braided stream conglomerates on the midfan, and laminated sandstones at the fan toe and on the basin plain. Significantly, debris flow conglomerates occur, for the most part, only near the source area, whereas braided stream deposits display a progradational pattern over laminated sandstones. This possibly represents the phase in fan evolution where a dynamic equilibrium has been established and reworked proximal sediments are being distributed downfan by intrafan channels.

The abstraction stage of source area evolution is represented by the retrogradational pattern shown at the top of the sequence. There appears to be a rapid transition from proximal facies to distal ones suggesting the rapid onset of this phase. Apparently, reduction of slope gradients in the source area was combined with rapid expansion of valley floors that provided storage for coarse sediments.

It is also interesting to note the periodic extension of coarse facies, especially debris flow conglomerates, into fan toe and basin plain environments. These may represent periods of fan head entrenchment when coarse sediment bypassed upper and mid fan segments to be deposited farther downslope.

The fans analyzed thus far have all developed in climates that have been interpreted as arid. Similar sequences can be found in humid region fans, however, and among the best analyzed of these are the alluvial fan sequences in the Stephanian coal fields of northern Spain (Heward, 1978).

The most proximal facies in these fans consist of those forming in valley and fanhead canyon settings. They consist of coarse unstratified conglomerates, trough cross-bedded sandstone and root-mottled mudstone (Fig. 4.26), and were probably deposited by a variety of processes including mass movement by debris flows and colluviation, stream flow, and suspension settling. The sediments are organized into fining-and thinning-upward cycles 15 to 65 m thick which are superimposed on thinning and fining upward megacycles more than 100 m thick. Heward (1978) interpreted the minor cycles to represent episodes of channel filling and abandonment, and the megasequences to represent cycles of uplift followed by slope reduction in the source area. In both cases, fan sedimentation responded to intrinsic factors in the source area and on the fan by forming regularly-developed sequences.

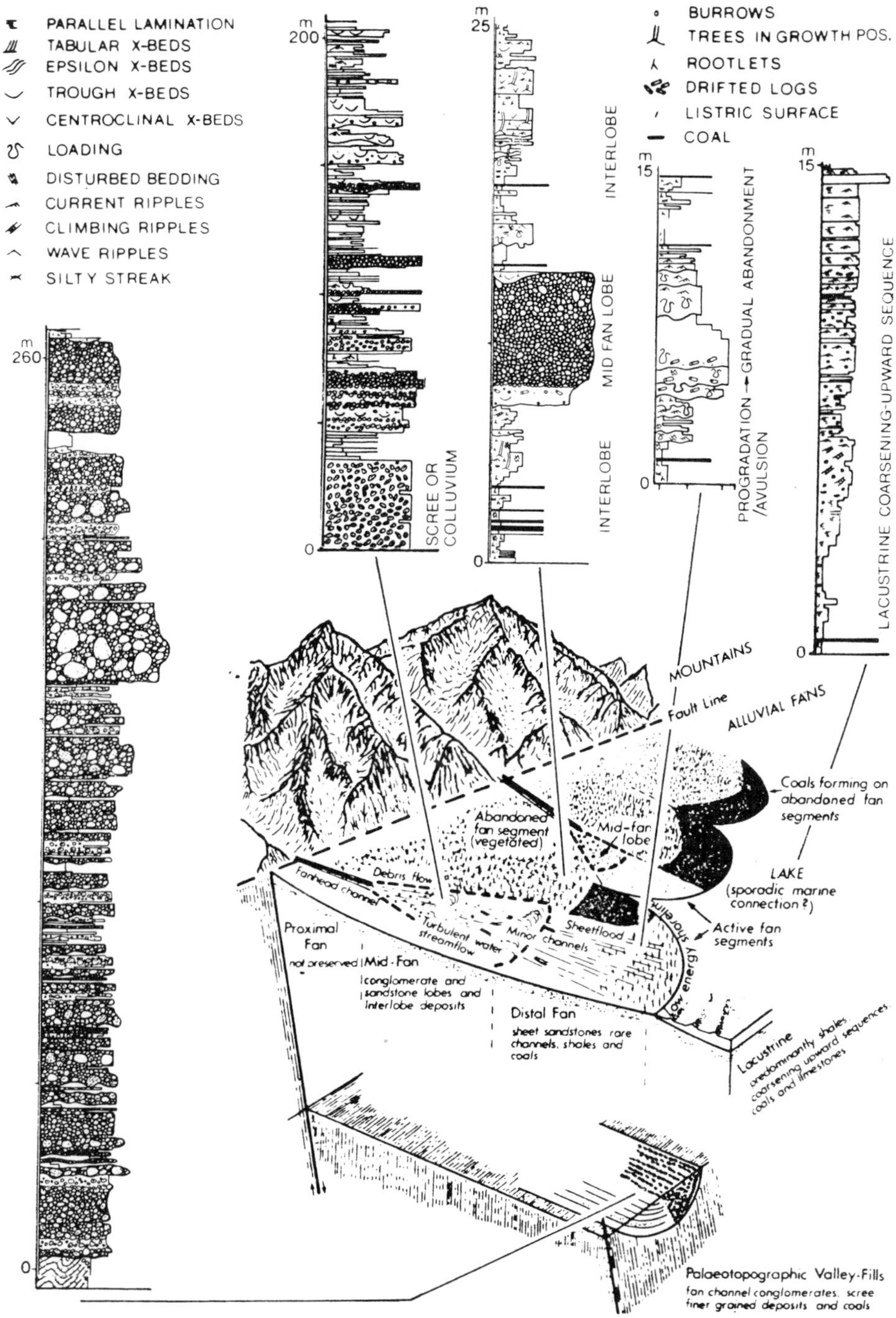

Figure 4.26 Vertical sections showing sequences in various parts of Carboniferous-age fans in Spain (after Heward, 1978a).

Mid-fan sediments consist of stratified and unstratified conglomerates and cross-bedded sandstones (Fig. 4.26). Individual conglomerate beds probably accumulated during major floods as debris flow or stream flood deposits with cross-bedded sandstones forming during waning flow of flood episodes.

Mid-fan sediments are organized into thickening-and coarsening-upward sequences superimposed on thinning-and fining upward ones. Heward (1978) interpreted the major cycles to have formed in response to slope reduction in the source area following episodes of uplift. The thickening- and coarsening-upward sequences accumulated during periods of lobe progradation caused by avulsion, but they might also represent alternating periods of fan head trenching. Minor storm cycles are also found in the mid fan where they consist of alternating coarser-and finer-grained layers superimposed on the longer period cycles.

Distal fan sediments consist of mudstones and fine sandstones. Sandstones were apparently deposited by sheetfloods during peak flood flows whereas interbedded mudstones and coals were deposited during intervening periods. These sediments occur in sequences up to 10 m thick that may thin-and fine-upward, or thicken-and coarsen upward. Sequences that thicken and coarsen upward are probably the distal equivalent of mid-fan lobe progradation caused by channel avulsion (Heward, 1978), or fan head trenching. Thinning-and fining-upward sequences, on the other hand were interpreted to represent lobe abandonment or climatic changes (Heward, 1978). However, they might also represent a phase of channel backfilling following a period of fan head entrenchment.

DISCUSSION

The question certainly arises at this point as to whether or not a fan model can be constructed, given the large number of interactive extrinsic and intrinsic controls that can affect fan evolution, without simplifying the situation so greatly as to render the model useless. We believe that it is possible to construct such a model realizing that it falls prey to all of the factors that inhibit the direct application of any model. Because the fan is so dependent on maintenance of a delicate balance between sediment delivery and removal, however, the effect of extrinsic factors is magnified and the propensity of complete sequences to form is greatly reduced. It may be that thick sections of fan sequences can be characterized by the presence of incomplete sequences, rather than being noted for the occurrence of complete ones.

If fan development begins with uplift of a relatively smooth surface, the resultant topography will consist of a low-relief basin that is separated from an upland of low relief by steep fault-controlled slopes. The basin floor would extend very near the marginal slopes so that it would be in a position to receive sediments of two radically different types (Fig. 4.27).

The steep marginal slopes would supply coarse, poorly sorted sediments that would accumulate as scree, colluvium, and debris flow deposits. Streams issuing from the upland would decellerate rapidly at the basin floor because of reduced gradients and rapid flow expansion. Initial sediment yield would be relatively coarse-grained because the streams would possess steep gradients in the source area, and the coarse sediment they carried would be deposited very near the basin margin (Fig. 4.27), leaving the finer grained sediments to be deposited on the basin floor away from the margin. The basin floor, however would extend very close to the basin margin, and during initial stages, therefore, sediments characteristic of proximal processes would be deposited in close association with sediments characteristic of more distal ones.

Development of the drainage network in the source area would proceed rapidly, and have an important effect on fan development. First, as the drainage basin expanded and the sediment yield increased, the fan would expand into the basin at the expense of the basin plain so that the basin floor would no longer be closely juxtaposed with the margins. Secondly, the sediments delivered to the fan would become increasingly coarse grained as source area slopes steepened. Nearly all of the erosional products of the source area would be delivered to the fan because the valleys at this time would lack storage capacity.

The drainage basin would very rapidly gain the ability to store coarse sediment on the valley floor of the trunk stream. Sediment yield to the fan would decline in quantity and caliber and basinal facies would expand onto the fan surface. This would be only a brief phase in fan evolution, and once the drainage basin entered a prolonged phase of extension, sediment yield would increase in response to increased discharge in the trunk stream and an increase in total area occupied by valley slopes.

During the stage of extension the fan would expand very rapidly and establish a thickening, widening wedge of material against the margin of the basin (Fig. 4.27). As the wedge increased in size its gradient would decline and this would permit differentiation of the fan into upper, middle, and distal segments and provide a more gradual transition of fan sediments into basin plain ones.

The rapid expansion would also cause lateral migration of fan facies and development of coarsening-and thickening upward sequences. The major fan process at this time would be expansion, and the internal organization of the fan would be subsidiary to growth.

Expansion of the drainage basin in the source area would slow as the stage of maximum extension was approached. Fan expansion would also slow and a stage of dynamic equilibrium would be approached as the gradient on the fan became accordant with that of the trunk stream in the source area. The valleys of the trunk stream and its major tributaries would be wide enough and their gradients low enough to significantly reduce the caliber of the sediments delivered to the fan. Those sediments supplied to the fan would be subjected to multiple episodes of reworking as the fan attempted to maintain an equilibrium profile in the face of intrinsic stress on the system.

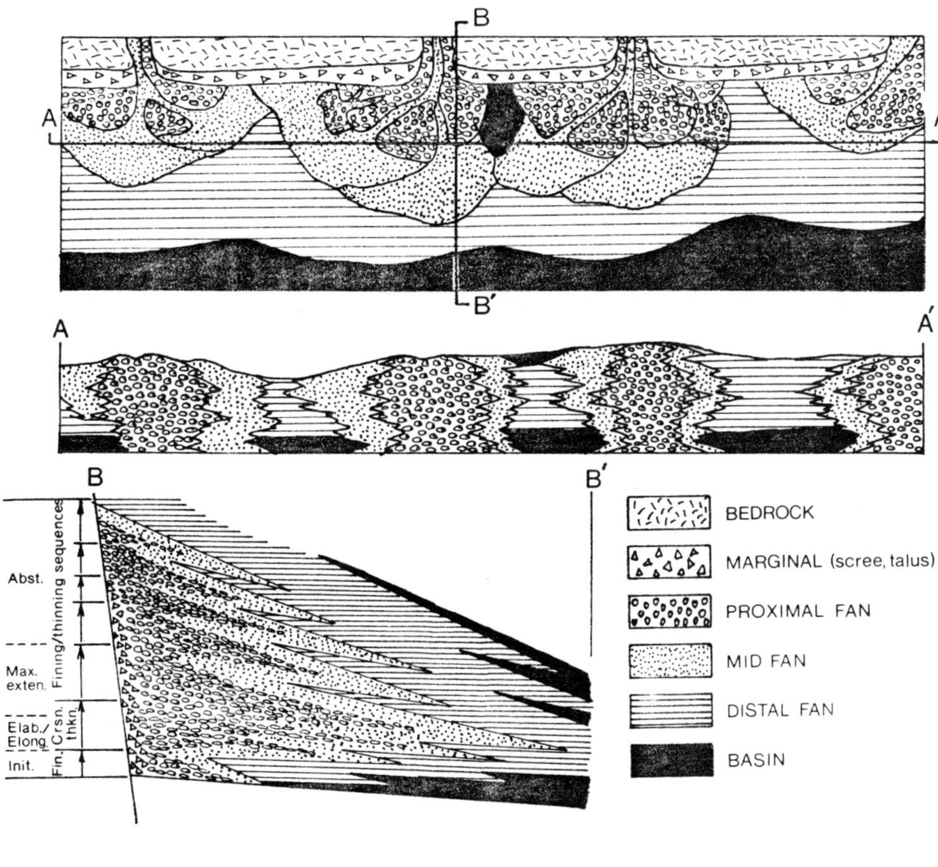

Figure 4.27 Diagrammatic plan view and cross-sections that would develop during evolution of a fan under the influence of intrinsic factors after a single episode of uplift.

It is during this stage that fan deposits would become most highly organized into minor sequences. Episodes of fan head entrenchment would deliver reworked sediment to the midfan where they would be deposited in coarsening/fining-upward sequences that would have their lateral, albeit finer-grained, equivalents in the lower fan and basin plain. Deposits of this phase, therefore, would characteristically be organized into minor coarsening-upward cycles, but the overall coarsening/fining-upward trend, established during the earlier phase of fan expansion, would be subdued because feedback mechanisms would act to stabilize the lateral positions of the various fan facies.

Eventually, sediment yield from the source area would change as slope gradients and stream gradients declined and storage capacity increased. Total quantity would be reduced, but clast size would decline even more rapidly because even the relatively small amount of coarse sediment still being produced would be preferentially stored in source area valleys.

As the stage of abstraction progressed in the source area, the sediment yield would continue to decline and fan area would contract. A fining upward trend would be established at the top of the sequence as fan facies migrated sourceward and sediment caliber from the source area was reduced. The rate of aggradation would slow reducing the intrinsic stresses on the system so that minor cycles would be less evident as the fan entered the final stages of evolution. The resulting deposit would be strongly asymmetrical consisting of a relatively thin sequence of sediments deposited during fan progradation and stabilization, and a thick one deposited during retrogradation (Fig. 4.27).

It is difficult at this time to adequately test if this model is an accurate reflection of reality because relatively few studies have been completed with the attention to detail necessary to establish the various stages of evolution. There are some indications, however, that this model can be applied to fan sequences. Steel (1974) for instance, measured several fan sequences in the New Red Sandstone. In at least two of them a sequence can be seen that consists of initially very coarse sediment at the base deposited primarily by debris flows, followed by a section of interbedded coarse and fine sediments showing no overall grain-size trend, and capped by a fining-upward sequence.

5

Effects of Extrinsic Stress on Evolution of Alluvial-Fan Sequences

INTRODUCTION

Thus far, we have analyzed the sequence of events that occurs during the evolution of a fan system following uplift of a smooth surface. Discussion has centered on the role that various intrinsic geomorphic factors have had on sediment yield to the fan and on distribution of sediments on the fan. These factors combine to produce a depositional sequence that can be considered the basic motif of fan architecture. In this chapter we consider how external stresses on the system modify the way intrinsic factors operate to produce variations in the basic motif that is recorded in stacked fan sequences.

The geomorphic principle that describes these interactions is termed complex response. A first-order response of a system to external stress may be predictable. Typically, for instance, a change in climate toward increased rainfall in the source area will increase runoff and sediment yield and cause aggradation on the fan. This response to external stress, however, may produce internal stresses that trigger additional responses. As the fan aggrades, for example, it may actually reduce sediment yield for a time by raising the base level of source area streams and causing them to aggrade and store sediment in their valleys. The stored sediment, however, is subject to remobilization if the fan aggrades to the point of instability and initiates a period of entrenchment that results in a second period of high sediment yield to the fan. In this case, the original change in external conditions, instead of producing the expected simple increase in sediment yield, has produced a complex response that has included at least two periods of increased sediment yield separated by one of reduced yield.

The major external stresses affecting alluvial fan systems are produced by climate and tectonism. Fan systems evolve to accommodate these stresses by changes in rates of erosion in the source area and aggradation of the fan surface. Changes in climate or renewed tectonism will change the way the systems evolve and thus the manner in which fan sequences accumulate.

TECTONISM

Experimental Results

We have already discussed the way in which a single period of uplift of a relatively smooth surface results in fan formation and evolution. In most situations, however, tectonism is periodically renewed after the initial episode, either because of unrelieved stress, or because of isostatic response caused by denudation of the uplifted area and loading of the basin. When this occurs in terrains containing previously established fan systems, a series of complex responses may be triggered that are controlled by several different feedback mechanisms.

We would expect, for instance, that lowering base level would rejuvenate streams in the source area and cause a rapid increase in sediment yield to the fan. Parker (1976, 1977) observed that lowering the base level on an experimental fan did, in fact, initially increase yield, but this was followed by a general decline in channel erosion in the source area and sediment yield to the fan (Fig. 5.1). The decline was not steady, however, but instead was interrupted by secondary episodes of large sediment yield. As base level was subsequently lowered during the experiment, slopes in the drainage basin steepened with each episode and sediment yield increased progressively, but in each case, secondary peaks persisted (Fig. 5.2).

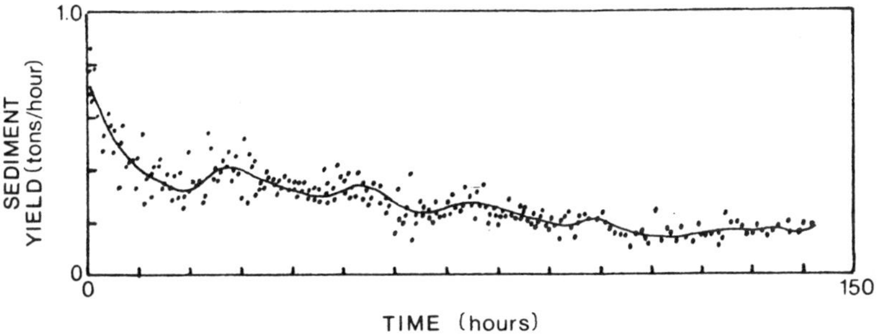

Figure 5.1 Sediment yield variations following stream rejuvenation during experimental drainage basin evolution (after Parker, 1976).

The occurrence of these secondary peaks probably illustrates the complex response of the drainage basin to base level lowering. After the initial period of uplift, a drainage basin evolves to the point of maximum extension when sediment production from steep interfluves is large. By this time, however, main valleys have increased their width and decreased their gradients to the point where they are able to store much of this sediment, especially the coarser grades, and sediment yield to the fan actually begins to decline in both quantity and caliber.

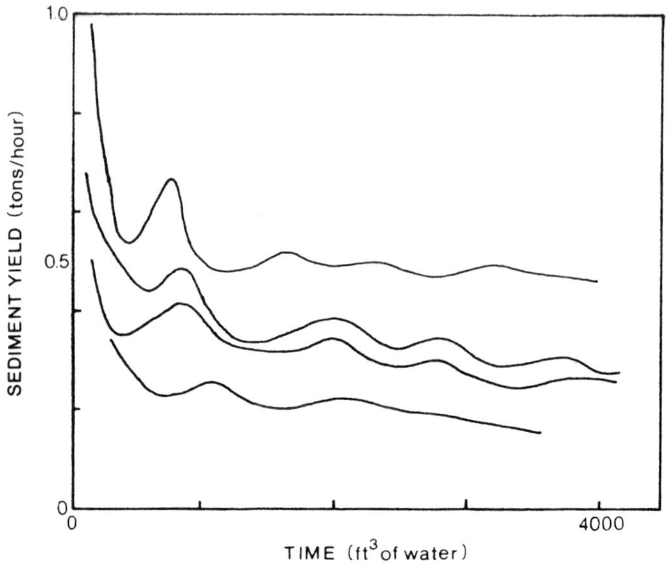

Figure 5.2 Sediment yield in an experimental drainage basin following four episodes of stream rejuvenation that produced progressively increasing relief (after Parker, 1976).

During the initial phase of rejuvenation following the lowering of base level, sediment stored in the valley of the trunk stream is eroded as the stream attempts to restore an equilibrium profile. An initial surge of coarse sediment is delivered to the fan, but the yield declines as the rate of channel erosion slows. As this occurs, however, the major tributaries are, in turn, rejuvenated. They increase their yield of sediment to the trunk stream and aggradation occurs as the stream steepens its gradient in order to accommodate the increased load. Sediment yield to the fan declines even more as this occurs until the channel gradient reaches a point of instability and the trunk stream begins to entrench its valley again and deliver a secondary surge of sediment to the fan. The sequence produced at the fan head by this process would consist of a first order fining-upward cycle composed of a series of second-order fining-upward cycles (Fig. 5.3).

The fan-head channel periodically entrenches itself during this period, both in response to its own threshold parameters, and to the cyclic changes in channel gradient occurring in the source area. Periodic fan-head entrenchment delivers cyclic pulses of sediment to mid-and distal fan areas that forms prograding depositional lobes and result in the accumulation of coarsening-upward and then fining-upward sequences lower on the fan.

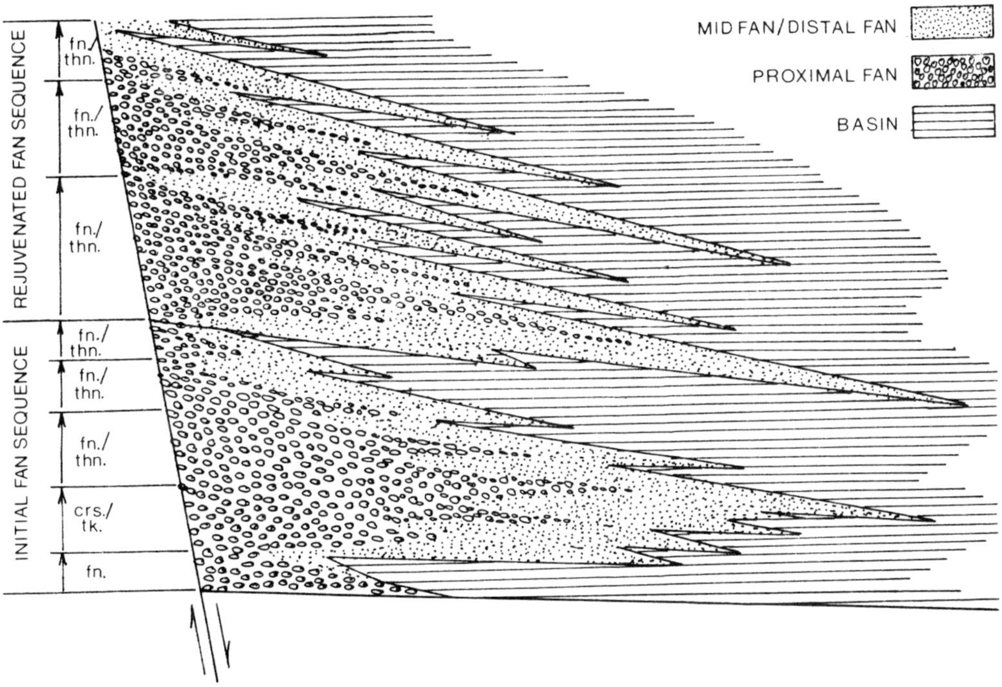

Figure 5.3 Cross-section through a fan showing the vertical sequence that would develop during a single episode of stream rejuvenation in the source area. Complex response to rejuvenation would deliver successively smaller and finer-grained pulses of sediment to the fan as progressively higher-order streams in the source area were rejuvenated.

Tectonic Controls on Fan Morphology and Sedimentation

The rate of tectonic uplift may have an important effect on fan evolution by superimposing constraints on the operation of the intrinsic mechanisms triggered by the system as it moves to restore the equilibrium profiles disturbed by the tectonism. Bull (1977), for instance, describes two modes of fan growth controlled by greatly different rates of tectonic uplift (Fig. 2.6). When the rate of uplift is very large, channel downcutting in the basin and deposition at the mountain front is promoted, and because the fan head is aggrading, the tendency for entrenchment to occur is reduced. As long as the rate of uplift exceeds the rate of downcutting of the trunk stream and the rate of aggradation of the fan head, entrenchment will not occur. Sediments will continue to accumulate adjacent to the mountain front as the system attempts to restore accordant gradients of the trunk stream and fan surface.

Through this process, segmented fans, consisting of concentric bands of approximately uniform slope, form around the fan apex. After each episode of uplift, a slope gradient, steeper than that of the remainder of the fan, is formed at the apex, and each concentric band, therefore, represents the slope established by fan aggradation in response to a single tectonic episode.

Even though sediment transfer downslope from the fan apex would not take place through a fan trench in this case, sediment transfer could occur by other mechanisms. Secondary lobes may develop in front of channels cut into the fan apex (Denny, 1965). They consist of sediment reworked from the primary fan and thus tend to be better-sorted and finer-grained (Heward, 1978). The initial sediment yield to the minor lobe would be relatively coarse, but later yields would decline in quantity and caliber as gradients of the minor channels declined. In some cases, however, the minor channel may extend itself headward to the point where it captures the main channel and induces a phase of major lobe building at its terminus.

Episodic fan entrenchment, of a minor nature, may occur

in response to intrinsic stress or threshold mechanisms and result in the accumulation of depositional lobes below the intersection point. Even these minor episodes of lobe progradation would probably produce deposits that become finer grained with increasing distance from the channel terminus. Initial sediments delivered to the lobe would be coarse grained, but as a new equilibrium profile was established in the channel, sediment yield would decrease in amount and size.

The rate of uplift in the second case is slow compared to the rate of channel cutting and fan erosion. Channel gradients in the source area tend to decline through time, and if they aren't steepened by uplift, the main fan channel is forced to incise the fan head in order to maintain an accordant profile with the source area. Unlike episodes of entrenchment caused by intrinsic stress, entrenchment in this case is permanent and sediments tend to bypass the fan head entirely, leaving the surface subject to soil-forming processes.

Renewed tectonism may rejuvenate source area streams, imposing a cyclic pattern on sediment delivery to the fan, but these sediments would accumulate on the middle and lower reaches of the fan rather than at the apex. Segmented fans might also form in this way, but the segments would represent a reduction of gradient in the lower fan through progradation of the toe rather than steepening of the fan apex by aggradation. Fan segments formed by this process would consist of a major prograding depositional lobe that formed in response to increased sediment yield from the rejuvenated source area streams. Superimposed on the major pulse of sediment, however, would be minor pulses imposed by complex response to feedback mechanisms in the source area.

Minor depositional lobes could also form on the lower fan in front of gullies established on abandoned parts of the fan apex. These would be similar in aspect to those formed on fans where rates of uplift are faster and permanent entrenchment does not occur. Sediments of these minor fans would consist of relatively fine-grained sediment reworked from the primary fan. The secondary channel could capture the primary channel, however, inducing a rapid influx of coarse grained sediment from the fan apex and initiating formation of a new major lobe.

EFFECTS OF TECTONISM ON EVOLVING FAN SEQUENCES

Schumm (1977) identifies four scales of depositional cycles in alluvial fans that can be produced by erosional evolution of the source area. Primary cycles are related to denudation following uplift and produce sequences that coarsen upward in response to extension of the drainage basin in the source area, and then fine upward in response to sediment storage in valleys in the source area (Fig. 4.5).

Under normal conditions, the primary cycle is interrupted by episodes of renewed tectonism that cause rejuvenation of streams in the source area and produce pulses of sediment influx to the fan. The character of the sequences produced during these episodes varies in response to the nature of the tectonism and to relative position on the fan surface and would consist of second and third-order cycles.

When rate of uplift is large relative to fan aggradation, fan segmentation results (Fig. 5.4) and an active fan segment is constructed adjacent to the mountain front. The segment will receive initial influxes of coarse sediment that decline in quantity and caliber through time as the trunk stream in the source area is periodically aggraded by rejuvenated tributaries and then eroded. In proximal areas, this will produce a series of fining-upward/thinning-upward cycles that become progressively finer-grained and thinner upward (Fig. 5.5). In the distal parts of the segment, however, the effects of episodic progradation and retrogradation will be more pronounced allowing for organization of these pulses into coarsening/thickening-upward and then fining/thinning-upward cycles. On the rest of the fan, progressive onlap will occur and the sediment size being supplied will decline, resulting in an overall fining-upward sequence (Fig. 5.6).

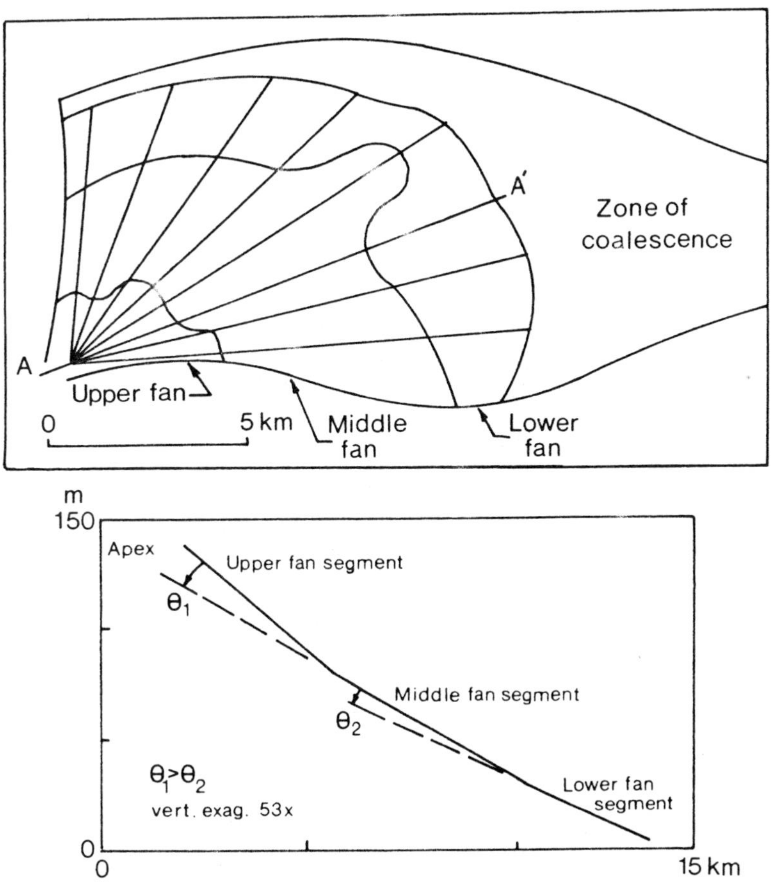

Figure 5.4 Morphology of a segmented fan showing fan segments in plan view (upper) and in longitudinal profile (lower) (after Bull, 1977).

When the rate of uplift is slow compared to fan aggradation, entrenchment of the fan head may occur. If this occurs, proximal parts of the fan are abandoned and soils form on the surface, but a depositional lobe forms on the mid-and lower parts of the fan (Fig. 5.6). Near the intersection point, the lobe may consist of a series of fining/thinning-upward cycles that in turn fine and thin upward in response to autocyclic aggradation and erosion in the trunk stream of the source area. Lower on the lobe, coarsening/thickening-upward and then thinning/fining-upward cycles will accumulate in response to lobe progradation (Fig. 5.6).

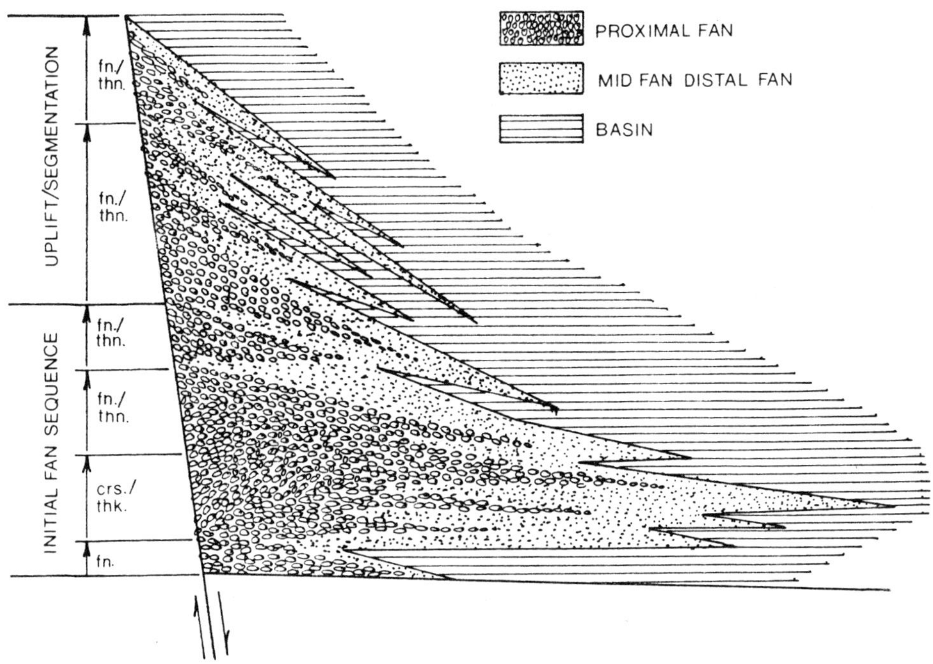

Figure 5.5 Cross-section through a fan showing the vertical sequence that would develop under conditions of rapid uplift and fan segmentation. Segments formed subsequent to uplift would show steeper profiles and smaller radii than their primary fans and initial sediment yield to the segment would consist of relatively coarse sediment from rejuvenated trunk streams in source areas.

Secondary fans may also develop in front of channels cut in abandoned fan surfaces (Fig. 5.7). They would consist of sediment reworked from the primary fan and they would be better-sorted and finer-grained than those sediments. Their deposits would tend to be thin, but they would otherwise resemble depositional lobes formed in front of major fan channels (Fig. 5.7).

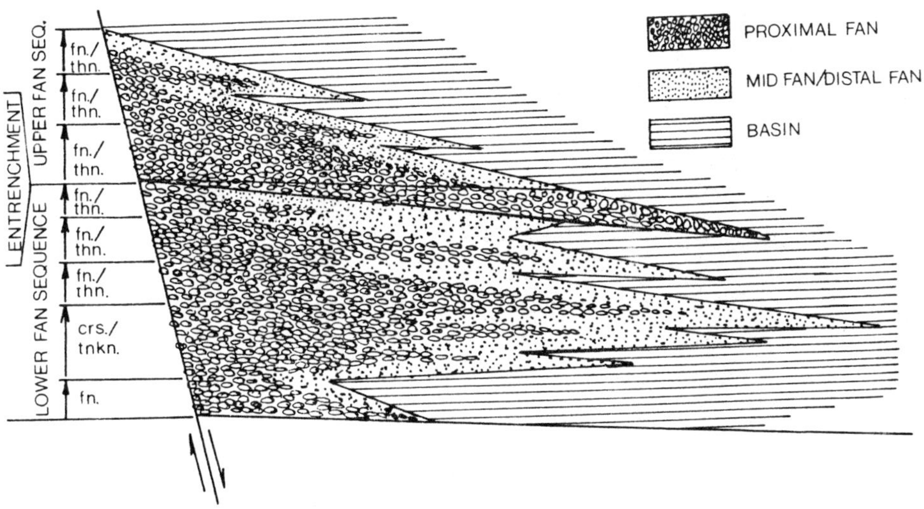

Figure 5.6 Cross-section through a fan showing the vertical sequence that would develop under conditions of slow uplift and fan entrenchment. When fan heads are entrenched, depositional lobes form where coarse sediments are delivered to mid-fan and distal fan regions.

Such channels, however, may migrate headward and capture the feeder channel of the primary fan. This would result in progradation of a major depositional lobe over the secondary fan. It would be reflected in a vertical sequence where the finer-grained sediments of the secondary fan are overlain by coarser-grained sediments derived from the source area (Fig. 5.6). Stream capture also results in abandonment of the former depositional lobe. In such cases the sequence could be truncated at any stage in development and a soil profile would develop on the inactive surface.

Fan entrenchment need not develop only in response to tectonism, but may be an intrinsic part of fan evolution that occurs when sediment yield from the source area begins to decline. This may occur when episodes of tectonism are widely spaced and equilibrium conditions are allowed to develop.

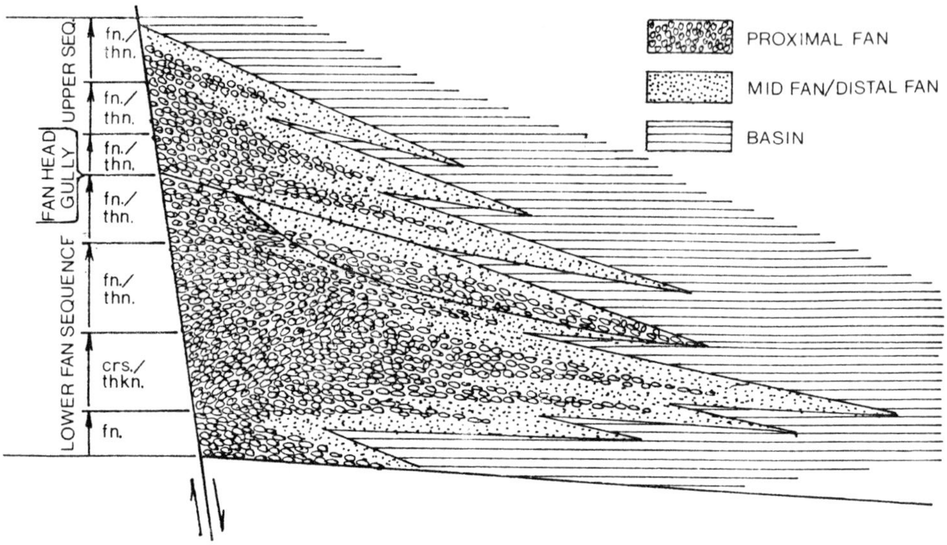

Figure 5.7 Cross-section through a fan showing the sequence that would develop under conditions of slow uplift and gully formation. Relatively fine-grained depositional lobes would form from reworked fan sediments at the termini of such gullies.

Depositional lobes will also form when this occurs, but each episode of entrenchment will produce a single prograding sequence that will thin and fine upward near the intersection point and coarsen and thicken upward, then fine and thin upward in more distal areas. Intrinsic controls may produce subsequent episodes of entrenchment, but these will not display the overall fining and thinning upward sequence that normally develops in response to tectonism, because complex response in the source area would not be involved.

Fourth-order cycles form in response to episodes of major flooding or to seasonality of hydrologic events. They consist of sequences that thin and fine-upward in response to conditions of waning flow, and they may occur anywhere within a fan sequence because they are produced by extrinsic factors that operate independently of the fan. Similarly, second and third-order cycles produced by tectonic events may also occur randomly in the major sequence because they are part of the extrinsic environment of the fan. Second- and third-order sequences produced by intrinsic factors, however, are triggered by threshold events that normally occur only after a fan has reached a state in equilibrium with the source area. They normally will occur only after the initial stages of fan aggradation and progradation, and will occupy the middle part of first-order cycles.

All of these cycles are superimposed on major cycles, 1000 m or more thick, termed "basin-fill" sequences by Heward (1978b). These form in response to episodes of basin development and record the initiation, filling, and eventual extinction of a basin.

In general, these sequences fine upward in response to gradual decrease in tectonic activity and progressive erosion of the source area, or, in some cases, expansion of the basin by back-faulting (Fig. 5.8). Either case results in the delivery of progressively finer-grained sediment to the fan and the expansion of basinal facies marginward. A fining-upward sequence is deposited and it is interrupted only in the case where inner fault systems are reactivated or major uplift of the margin recurs with the return to a more active tectonic regime.

Tectonic Controls on Architecture of Fan Sequences

The association of alluvial fans and tectonically active terrains is axiomatic. Uplifted source areas juxtaposed laterally with depositional basins are ideal locations for accumulation of coarse sediment wedges against fault or fold-controlled scarps. In general, the sedimentary processes active during the accumulation of these wedges vary only in response to local variations in climate, source area composition and drainage basin characteristics, and regional gradient. As a result, the basic motif of fan sedimentation, described in Chapter 4 and earlier in this chapter, does not change significantly in response to variations in tectonic style. The tectonic regime in which alluvial fans occur does, however, affect the way in which the basic motif is repeated in basin-filling sequences as well as the geometric relationships of the packages of sediments deposited during fan-building episodes over the life of a basin.

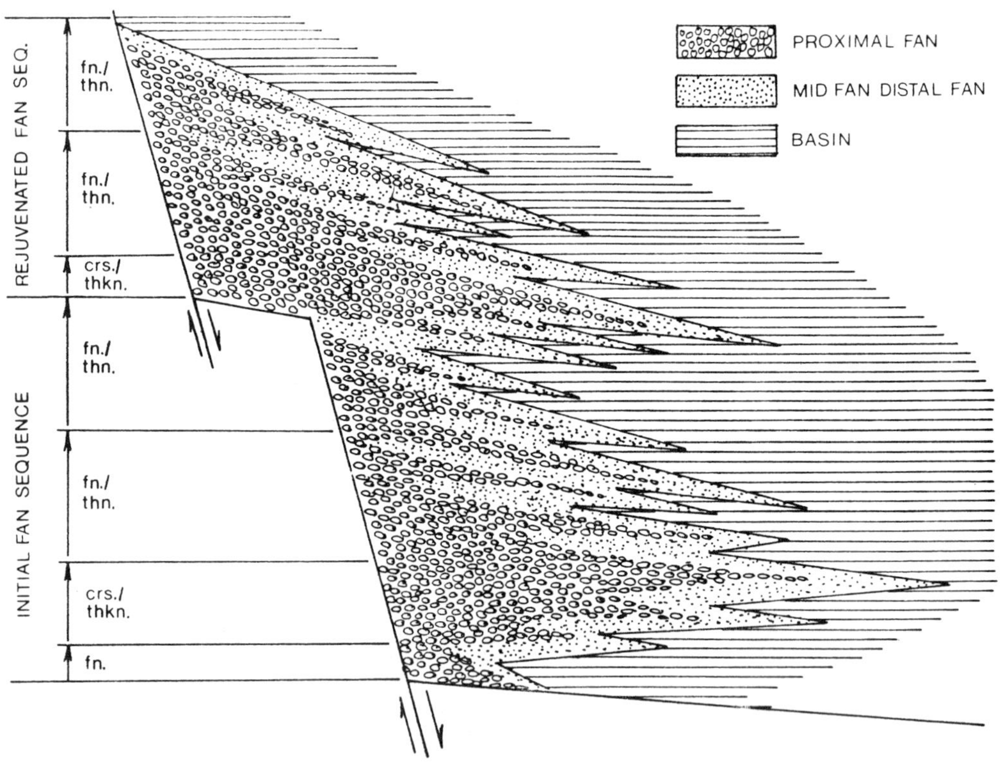

Figure 5.8 Cross-section through a fan showing the sequence that would develop where backfaulting occurred and source areas retreated from the basin. Rejuvenated fan sequences would be relatively coarse grained, but because the source area retreated, basin-filling sequences would fine upward.

Tectonism results from tensional, compressional, or transcurrent stress. Tensional stress produces grabens bounded by high-angle normal faults, and compressional stress generates folds, high-angle reverse faults and thrust faults. Transform faults are the result of transcurrent stress. The dominant movement along a straight transform fault is lateral motion along a high-angle fault, but where the fault trace curves, a complex regime of tensional and compressional stresses is established that can produce dilational basins and uplifted source areas (Crowell, 1974).

The simplest basin margin sequences occur where alluvial fans prograde into basins bounded by single normal faults. The basic fan motif may evolve to completion, but most often incomplete and complex sequences result from interruptions in fan evolution by episodes of basin subsidence during which source areas are uplifted, fans become segmented, and fan facies migrate basinward. Intervening periods of quiescence along the fault allow drainage basins to evolve to abstraction resulting in decreased sediment yield to the fan and sourceward migration of fan facies.

The resulting megasequence that develops along a single normal fault, therefore, may consist of both individual coarsening-upward and fining-upward fan sequences, but along margins of continuous uplift or subsidence, sequences with no pronounced grain-size trends may accumulate (Wilson, 1980) (Fig. 5.9).

Grabens may also be bounded by multiple normal faults that step down into the basin. The faults tend to become progressively younger away from the basin resulting in the migration of the source area away from the basin. In such a situation, a fining-upward megasequence develops, but minor sequences formed during episodes of fan progradation are commonly superimposed on the major cycle (Bluck, 1967; Belt, 1968; Sestini, 1970; Steel and Wilson, 1975) (Figs. 5.10, 5.11).

Even the results of simple vertical movements in grabens

however, may interact in a complex manner with other factors in the environment, especially those controlling rate of sediment yield to the basin, to produce a variety of basin-filling sequences. Miall (1978) relates rates of uplift

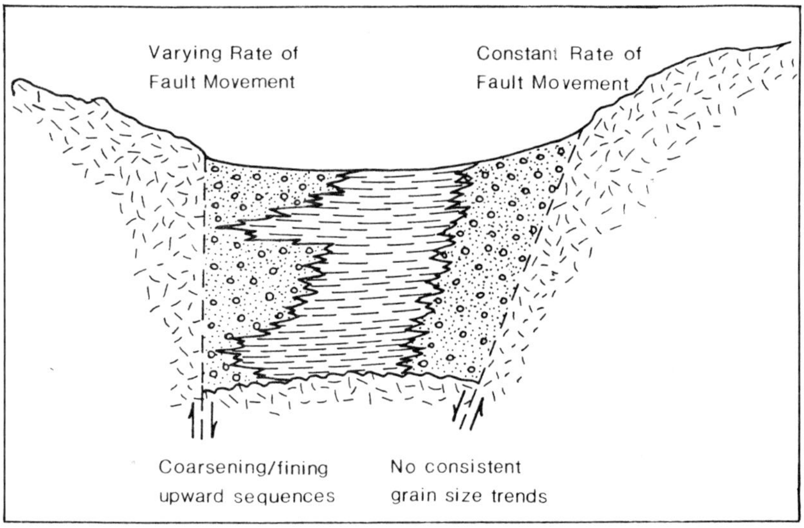

Figure 5.9 Types of basin-filling sequences developed along margins bounded by single faults with different rates of movement. Coarsening and then fining-upward sequences develop along faults with varying rates of movement whereas sequences with no consistent trends accumulate along faults with a constant rate of movement (after Wilson, 1980).

and downward erosion in the source area to rates of subsidence and vertical aggradation in the basin. He proposes five possible basin-filling sequences that are mainly dependent on the response of marginal fans to relative variations in the factors (Table 5.1).

Where compressional forces are active during basin filling, even more complex architectures may develop within basin margin sequences. Folding and reverse faulting at the margins may be expected under such conditions resulting in rotation of previously deposited sequences. In addition, the progressive basinward propagation of tectonic deformation causes further complexity by uplifting and

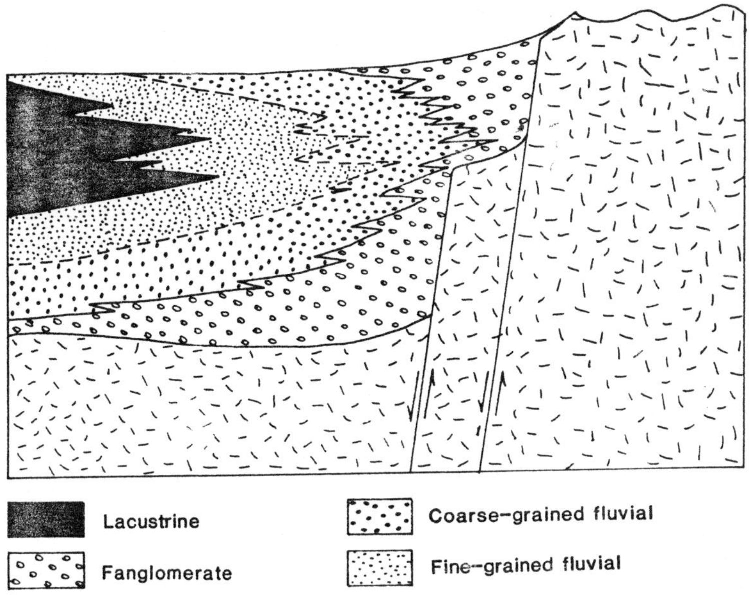

Figure 5.10 Diagrammatic cross-section through the margin of a typical basin of Carboniferous-age in northeastern North America showing the effects of back-faulting on marginal sedimentation (after Belt, 1968).

Table 5.1 Relationship between sedimentation and tectonism in a nonmarine sedimentary basin (after Miall, 1978)

Conditions	Geomorphic changes	Sediment changes	Facies belt behaviour
1. U=e,S=a	Equilibrium	Equilibrium	Position stable
2. U>e,S<a	Increased relief, slope, fan segmentation	Increased load and calibre, coarsening-up cycles	Basinward migration (progradation)
3. U>3,S≥a	Quasi-equilibrium	Coarsening-up cycles	Position probably stable
4. U<e, S<a	Reduced relief, headward stream erosion, fan entrenchment	Decreased load and calibre, fining-up cycles	Sourceward migration
5. U≤e,S>a	As in condition 4	Decreased load	Equilibrium or sourceward migration

Note: U = rate of uplift in source area; S = rate of subsidence; e = rate of downward erosion in source area; a = average rate of vertical accretion of proximal fan deposits (all rates relative).

rotating earlier deposited units and exposing them to erosion. Syndepositional unconformities are produced along the margins of the basin when this occurs as the uplifted deposits are cannibalized and serve as sources for younger sequences in the basin (Fig. 5.12). These unconformities extend only a few kilometers into the basin where deposition is continuous.

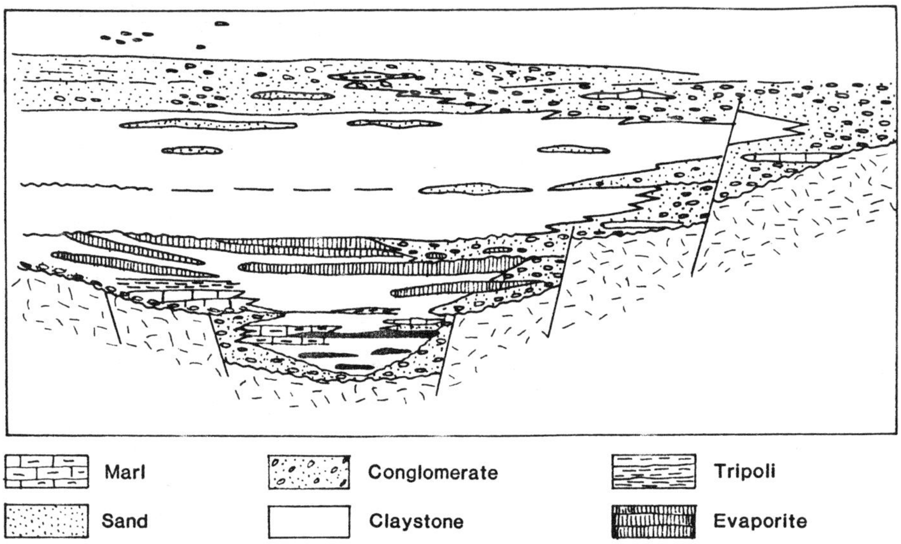

Figure 5.11 Cross section through the Miocene and Pliocene sediments of Tuscany (Italy) showing the effects of backfaulting on sediments in the basin (after Sestini, 1970).

The deposits in the basin tend to coarsen upward, partly in response to the basinward shift of deformation and also in part to accelerated uplift and erosion of source areas (Miall, 1978) (Fig. 5.13). Deposition in the basin tends to be continuous away from the axis of deformation so that the syndepositional unconformities being produced at the margins die out basinward.

Such basin margin unconformities have been termed syntectonic unconformities (Riba, 1976) and the degree of complexity that they introduce into basin-filling sequences increases as tectonism progresses. During initial stages of basin evolution, deformation may result simply in rotation of beds, thinning of marginal strata by erosion, and down-basin thickening of equivalent strata (Fig. 5.14). This stage is represented by the Tertiary Duchesne River Formation on the eastern margin of the Uinta Basin, western U.S.A. The formation thickens markedly into the basin and the poorly stratified conglomerates at the mountain front

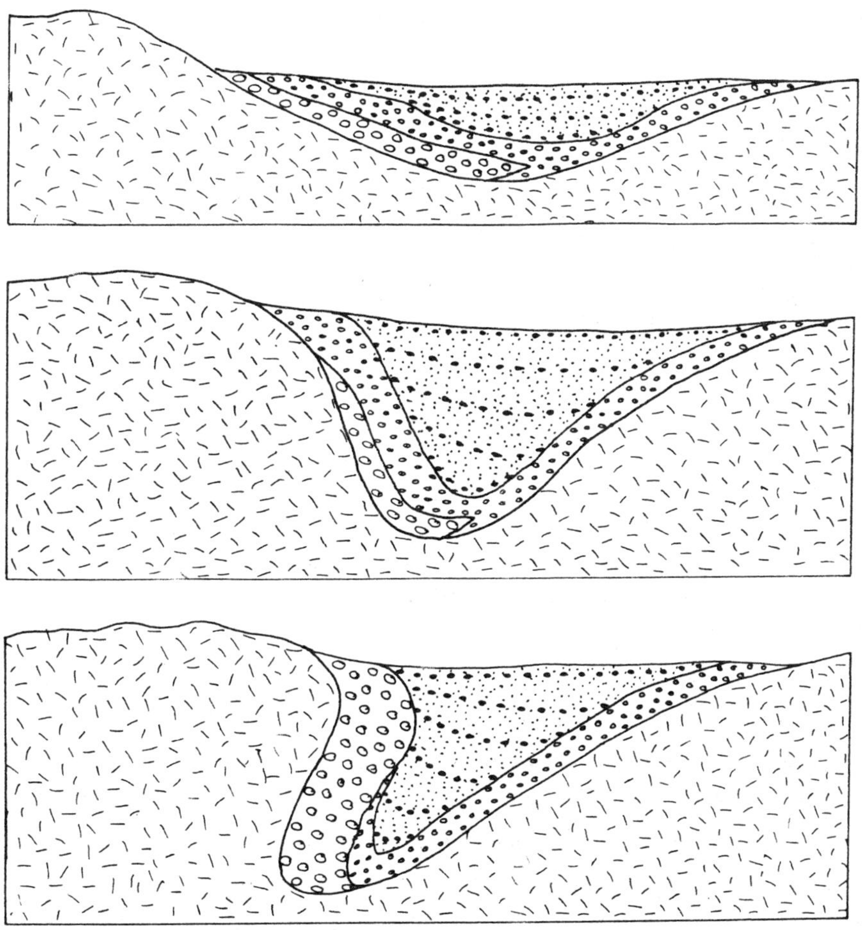

Figure 5.12 Three diagrammatic cross-sections through part of the Kvamshesten district of Norway showing progressive deformation of marginal sediments by thrusting of the northern margin to the south (after Bryhni and Skjerlie, 1975).

are interpreted to be lag deposits that accumulated slowly as finer-grained sediments were winnowed out and carried basinward (Andersen and Picard, 1974) (Fig. 5.15).

This pattern of sedimentation is one of sedimentary offlap, but if uplift ceases, offlap may be followed by onlap of the erosional surface (Riba, 1976) (Fig. 5.14). The angle of discordance between rotated beds and the onlap beds decreases basinward and a simple tectonic unconformity is formed. Such unconformities are found in the Ventura Basin of California (Baily and Johns, 1954) and in the Pyrenees of Spain (Riba, 1976).

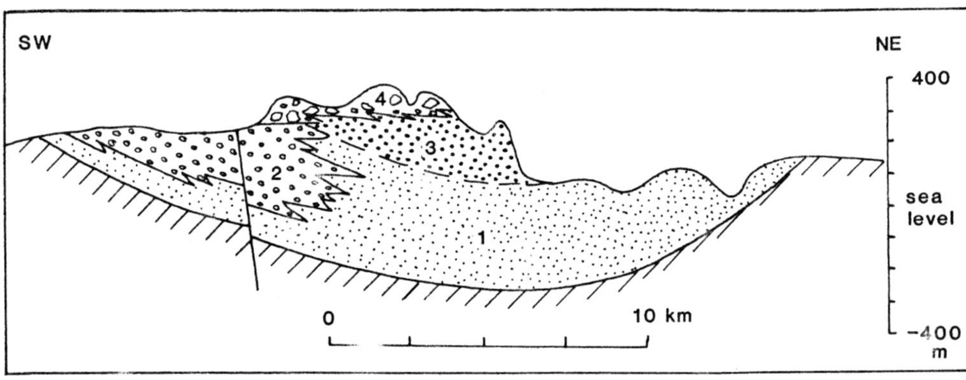

Figure 5.13 Cross-section through a coarsening-upward clastic sequence showing two conglomeratic wedges separated by a local unconformity representing a period of maximum rate of uplift (after Miall and Gibling, 1978).

The unconformity itself may be rotated and uplifted if tectonism is renewed resulting in a deformed syntectonic unconformity (Fig. 5.14). Further rotation of the onlap sequence, followed by formation of an erosional surface on the uplifted edge may produce yet another syntectonic unconformity and the whole package of basin margin sediments may represent a composite unconformity (Riba, 1976) (Fig. 5.14, 5.16).

Lateral movement dominates along transform faults. If the fault is long, straight, and vertical the opposing sides of the fault merely slide by one another (Crowell, 1974). If the fault is curved or non-vertical, however, compressional and tensional forces are established. These are relieved by complex vertical motions that form regions of uplift and subsidence along the fault. At double bends

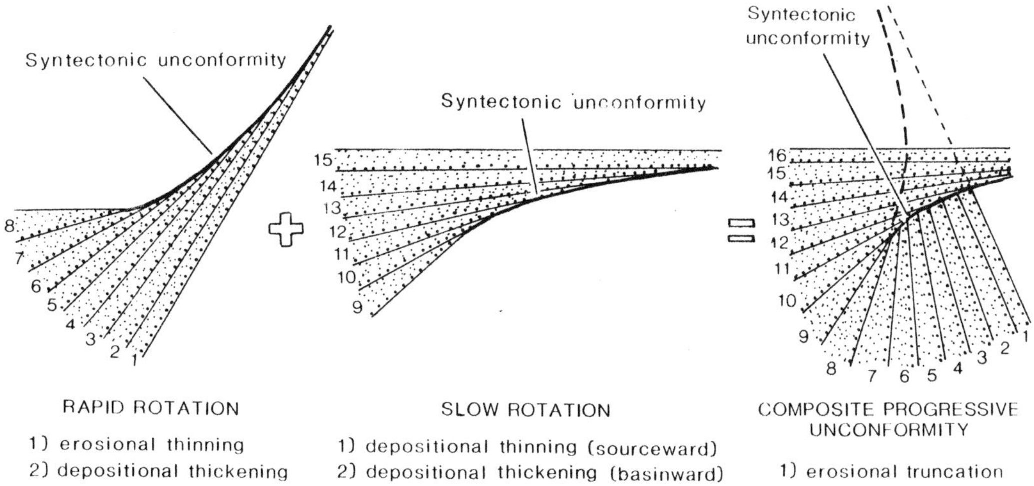

Figure 5.14 Stages in the formation of syntectonic unconformities showing how two simple syntectonic unconformities can combine to form a composite progressive unconformity (after Riba, 1976).

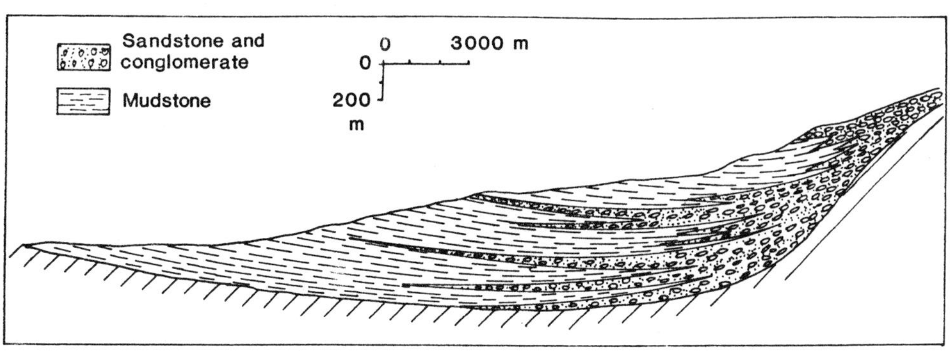

Figure 5.15 Cross-section through the Tertiary Duchesne River Formation of the Uinta Basin (U.S.A.) showing the initial stages of formation of a syntectonic unconformity by erosional thinning at the basin margin and basinward thickening where deposition was continuous (after Anderson and Picard, 1974).

in the fault, for instance, compression causes uplift at concavities in the curve while subsidence occurs at convexities where compression is produced (Fig. 5.17).

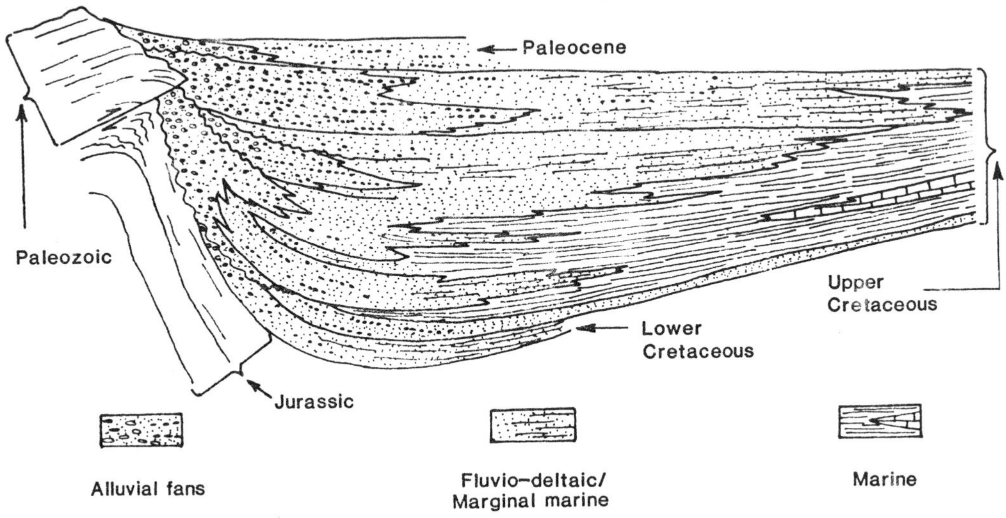

Figure 5.16 Cross-section through the Sevier orogenic belt showing composite syntectonic unconformities developed along the margin of the basin near the thrust block (after Armstrong, 1968).

The basins produced in such a setting tend to be asymmetric with one side bounded by a steep single fault and the other by a series of faults stepping down obliquely into the basin (Fig. 5.17). The Ridge Basin of southern California (USA) (Crowell and Link, 1982) and the Hornelen Basin of Norway (Steel, 1976; Steel, et al, 1977) apparently formed by such a mechanism, and although they differ markedly in the type of sediments that accumulated in the center of the basin, they show many similarities in the style of sedimentation along their margins.

Small fans consisting of very coarse grained sediment accumulate along the steep side of the basin. The Violin Breccia attains a thickness of 11 km along the San Gabriel Fault in the Ridge Basin (Crowell, 1982a) and up to 25 km of sediment was deposited in fans along the Nordfjord Fault in the Hornelen Basin (Steel, et al, 1977). In neither case, however, does the stratigraphic thickness represent the true vertical thickness of fan sediments in the basins.

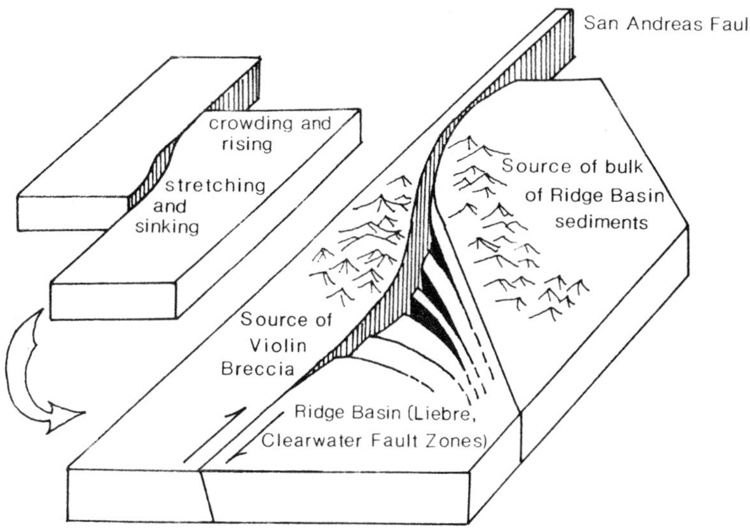

Figure 5.17 Block diagrams showing (a) distribution of stresses at curves in a transform fault; and (b) application of the concept to the formation and evolution of the Ridge Basin in southern California (U.S.A.) (after Crowell, 1974; and Crowell, 1982).

It is not uncommon for one side of a transform fault to remain fixed in place as the other slides by. When this occurs the same terrains continue to be uplifted as others are continuously depressed for long periods of time. Transtensional basins are formed that continuously widen and deepen through time, and they are bordered by fixed source areas experiencing continuous uplift in the region of compression (Crowell, 1974, 1982b) (Fig. 5.17). As depocenters move past source areas younger strata are deposited over older in an overlapping or shingle-like manner rather than in true superposition (Fig. 5.18). In this way, great stratigraphic thicknesses of sediment may be deposited in excess of the total subsidence occurring in the basin.

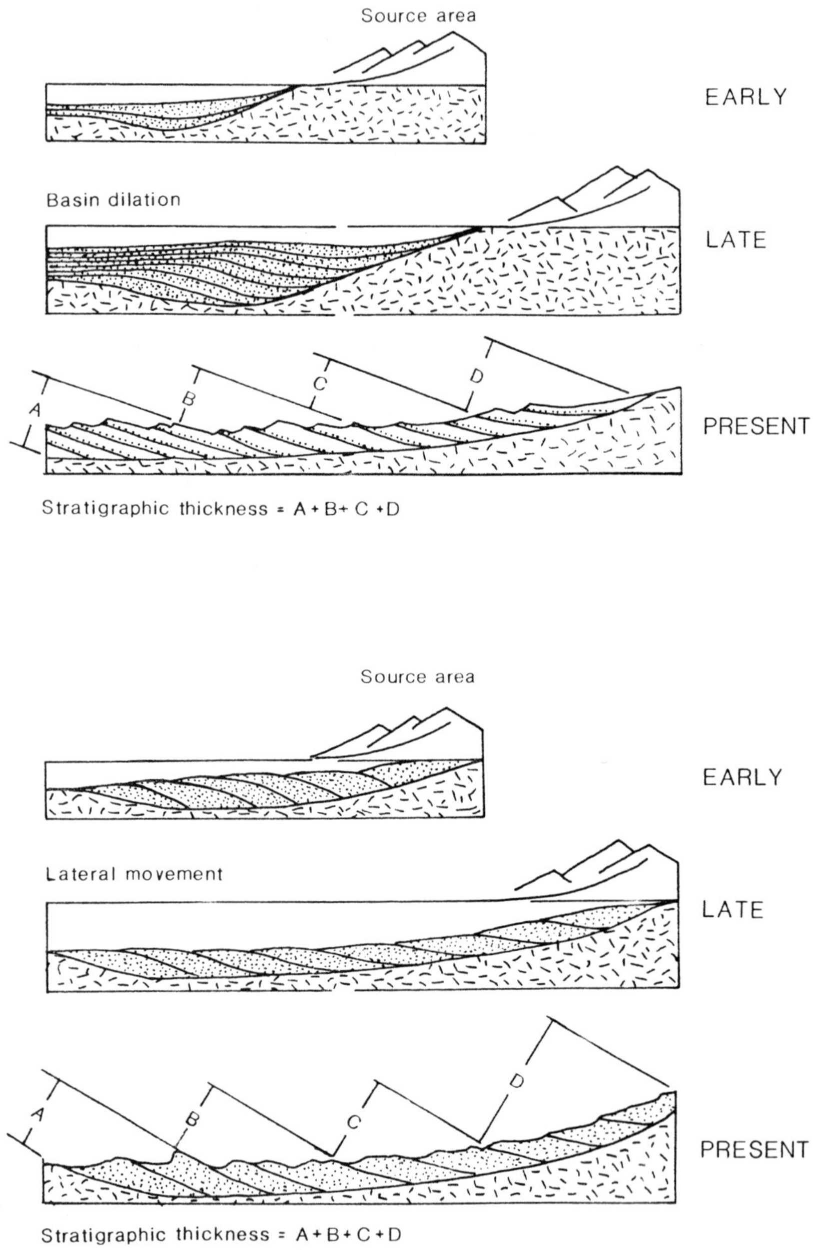

Figure 5.18 Diagrammatic cross-sections through a transtensional basin showing the geometric relationships as sediments fill a basin that dilates away from a source area (upper) and a basin that moves by a source area (lower) (after Crowell, 1982).

Finer grained fans of broader lateral extent were deposited along the more gently dipping side of both the Hornelen (south) and Ridge Basin (northeast). In these areas they were better organized into typical fan facies possibly in response to the gentler gradients on which they were deposited. The fans grade down dip into alluvial plain facies in the Hornelen Basin, but they grade downdip into a variety of marine and lacustrine shoreline facies in the Ridge Basin (Link, 1982, Link and Osborne, 1982). Like their counterparts on the steep sides of the basins, the fans on the more gently dipping side were supplied from a single source and are also arranged in an overlapping manner (Link, 1982) (Fig. 1.8).

On both sides of transtensional basins, fan sequences tend to coarsen upward in response to regularly recurring uplift in the source area and subsidence in the basin. Complete evolution of source area drainage basins is probably not possible under such conditions, and because the stage of abstraction is rarely, if ever, reached, the fining/thinning upward part of complete fan sequences is not a normal component of basin fill sequences in transtensional basins.

CLIMATE

Introduction

Climate has frequently been cited as a factor that affects fan morphology and sedimentation, and climatic change is commonly used to explain variations in fan sequences. The response of fans to climatic change is highly complex, however, and any interpretations of its effects must take into account the large number of variables affected by climatic change and the complex responses they may trigger. The situation is made even more complex by the probability that these responses were different during different periods of earth history.

Climatic Controls on Sediment Yield

It has become axiomatic that increased runoff and sediment yield are the direct result of increased precipitation. This statement is by no means completely true, however, and the part that is true is the result of a complex relationship dependent on several variables.

In general, when annual precipitation increases, annual runoff also increases. The amount of runoff declines for a given yearly amount of precipitation if temperatures rise, however (Fig. 5.19), so that the rate of increase in runoff slows with increasing mean annual temperature (Langbein, *et al*, 1949). This relationship occurs because the rate of evaporation increases with increasing temperature and reduces the amount of moisture available for runoff.

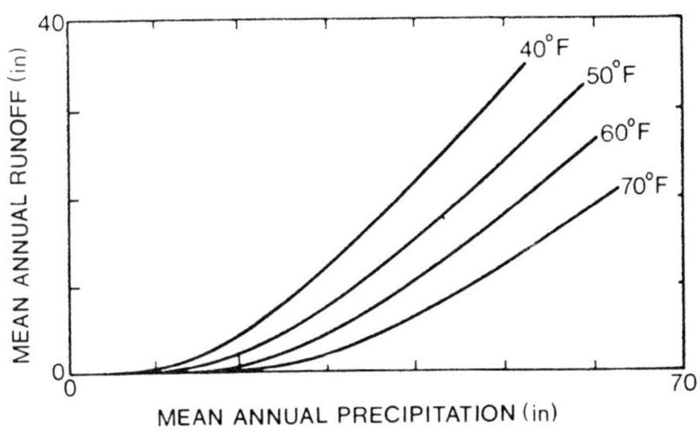

Figure 5.19 Effect of average temperature on the relationship between mean annual runoff and mean annual precipitation (after Langbein, et al, 1949).

The relationship between precipitation and sediment yield is also an indirect and complex one. In arid regions (<25 cm/year precipitation) an increase in precipitation from 0 to 25 cm/year increases the sediment yield. As precipitation increases, however, sediment yield may actually decline because the vegetative cover increases in response to greater availability of surface moisture (Langbein and Schumm, 1958) (Fig. 5.20).

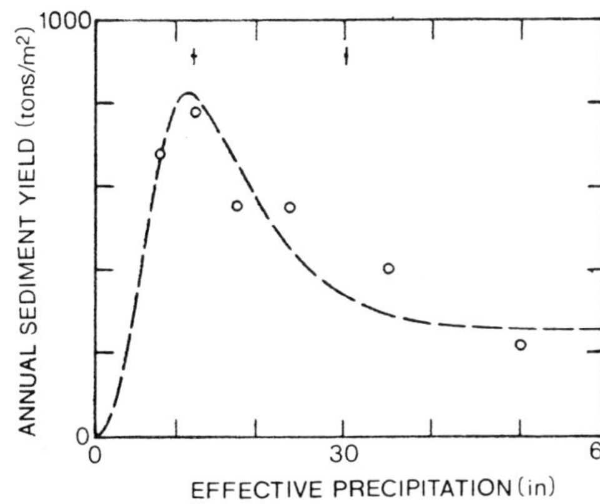

Figure 5.20 Variation in sediment yield in response to changes in effective precipitation (after Langbein and Schumm, 1958).

The relationship between runoff and temperature can be combined to predict the effect of increased precipitation on sediment yield for various temperatures (Fig. 5.21). Note that peak sediment production occurs at progressively greater values of mean annual precipitation as temperature increases (Schumm, 1965). Even this relatively simple relationship would almost certainly be complicated by changes in vegetational types that might occur with increased temperature and a change in type and amount of sediment made available for erosion under changing weathering regimes.

The foregoing analysis is based on the assumption that climatic change would occur slowly enough for changes in vegetational patterns to respond at the same rate. Knox (1972), however, suggests that changes in vegetation might not be able to keep pace with climatic change. If climate and vegetation changed at the same rate, an increase in precipitation in a semi-arid region might actually cause sediment yield to decline if the additional moisture resulted in an increase in the vegetational cover. If, however, the increased cover lagged behind the change in climate, the initial response might be a sudden increase in sediment yield followed by a rapid decline as the vegetational cover improved (Fig. 5.22). Such abrupt fluctuations in sediment yield to the fan would be difficult to distinguish from threshold responses to intrinsic stress such as fan-head trenching or channel avulsion. The effects of climatic change, however, would be widespread, whereas response to intrinsic stress would be limited to a single fan.

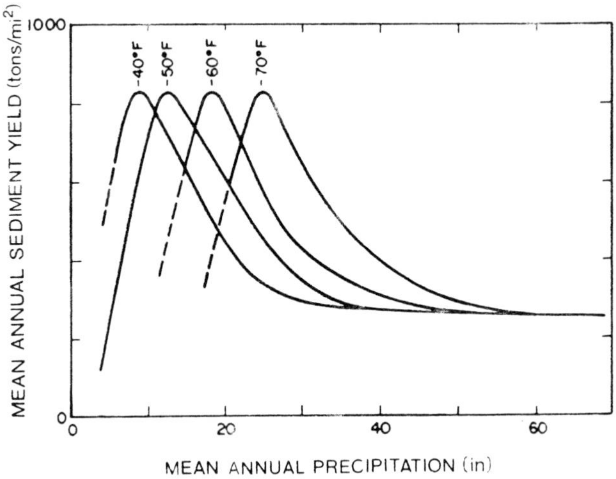

Figure 5.21 The effect of temperature on the relationship between mean annual sediment yield and mean annual precipitation (after Schumm, 1965).

Flood flows are the dominant events in fan sedimentation, but the way in which alluvial fan systems respond to these flows differs greatly between systems in arid or semi-arid regions and those in humid regions. Flood size decreases dramatically as mean annual precipitation increases over basins of similar size (Knox, 1972) (Fig. 5.23), for example, and the nature of the flows also changes. In arid and semi-arid climates, peak flows occur soon after storm peaks, and the bulk of the precipitation is included in the runoff. In humid climates, however, vegetative cover slows runoff and promotes storage of precipitation in the soil cover. Kittredge (1948), for instance, shows that surface runoff occurs only during intense storms in areas of complete vegetative cover, and even then only 3 percent of the precipitation may be involved in surface flow. Where slopes have been denuded, however, up to 60 percent of the precipitation is involved in runoff.

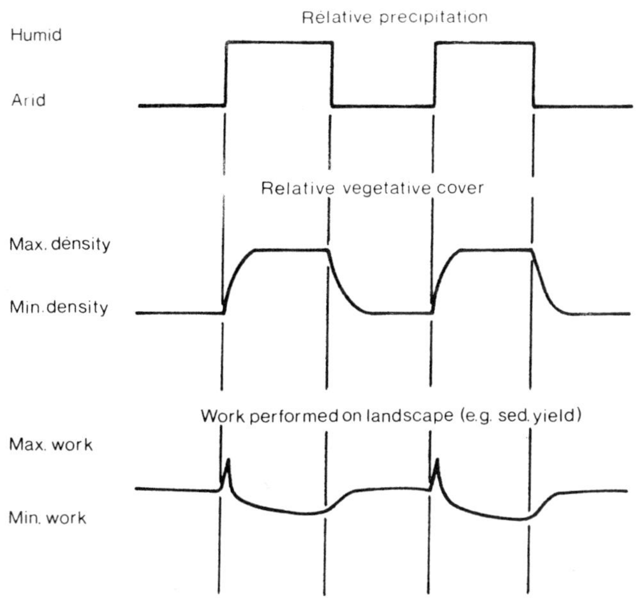

Figure 5.22 Vegetational and geomorphic response to abrupt climate change (after Knox, 1972).

Because of the nature of the response to storms, floods in humid regions reach peak flow more slowly and wane more slowly than in arid regions, and the magnitude of the flood is reduced. One result is that stream flow is more uniform in humid regions and deposits of individual flow events are less sharply defined.

Relation Between Vegetation and Sediment Yield Through Geologic Time

The preceding discussion has shown that the complexity of the effects of climate on fan evolution is due, in part

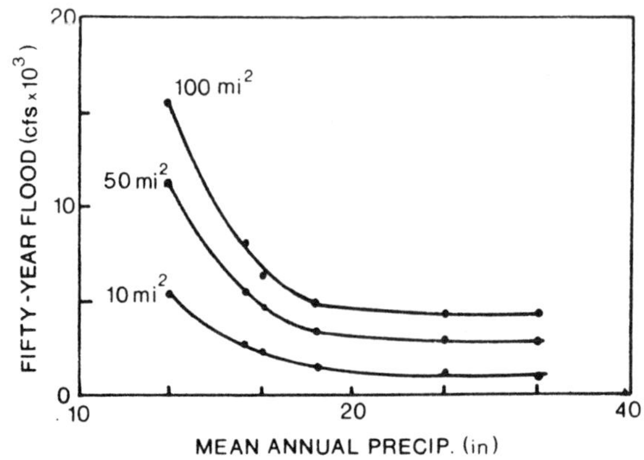

Figure 5.23 Relation between flood magnitude and mean annual precipitation for three sizes of drainage basins (after Knox, 1972).

at least, to the way vegetation affects runoff and sediment yield. The effects that vegetation have had on fan sedimentation have changed through time because the nature of the vegetative cover on land surfaces has changed as plants have evolved through geologic time (Schumm, 1977). In order to complete our analysis of the effects of climate on alluvial fan processes, we must now consider the way plant evolution has altered the way in which sediment is mobilized in the source area and delivered to the fan.

The evolution of terrestrial vegetation can be divided into four stages (Russel, 1956). The prevegetation stage spans most of geologic time through to the Silurian when significant colonization of the land by vascular plants occurred (Retallack, 1985). Through most of the early Paleozoic, however, plant forms were primitive and vegetation was confined to nearshore and humid coastal plain areas (Schumm, 1977). Streamside galleria forests probably occupied channel margins by the late Devonian and dry interfluves were apparently colonized by herbaceous vegetation by this time. Forestation of dry soils was probably achieved by the early Carboniferous, and well-drained, forested paleosols of early Triassic age have been described (Retallack, 1985).

Conifers evolved in the Jurassic and deciduous trees appeared in the Cretaceous, and these forms were capable of surviving seasonal, as well as longer period climatic fluctuations. Consequently, they were able to colonize upland interfluves. Finally, by the Miocene, grasses capable of forming thick vegetative mats on soils had evolved and runoff and erosion became similar to those of today.

During prevegetation times, runoff and erosion must have responded to precipitation much as they do in present-day arid regions. Episodes of runoff accompanied each storm period, hydrographs were peaked, and most of the rainfall was immediately transformed into runoff. Such responses, however, must have occurred in a similar way in all climates because losses to transpiration did not occur then. The only loss would occur through evaporation which would increase with increasing temperature (Schumm, 1977).

Runoff would have been directly related to precipitation for similar-sized drainage basins. Significant runoff, however, would not have been produced in those regions receiving less than 12 cm/year mean annual precipitation because of losses to evaporation and infiltration (Langbein, et al, 1949) (Fig. 5.24).

Sediment yield would have responded similarly to climatic change. Sediment yield would have increased with increasing mean annual precipitation to the point where very humid conditions would have been encountered. At that point, sediment yield would begin to decline as chemical weathering increased (Fig. 5.25) and as the density of runoff channels in headwaters decreased under the impact of very large amounts of rainfall.

Sediment delivery to the fan, however, might have been complicated by intrinsic mechanisms that would operate under arid or semi-arid conditions and impose a cyclic regularity to sediment influx (Schumm, 1977). Garner (1959) has shown that under semi-arid or arid conditions, sediments move off

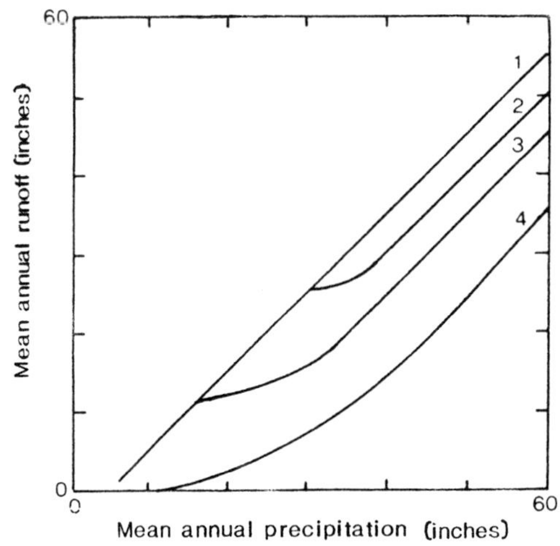

Figure 5.24 Hypothetical set of curves showing the relationship between precipitation and runoff during geologic time: (1) prevegetation; (2) prevegetation to Carboniferous; (3) Carboniferous to early Tertiary; (4) post early Tertiary (after Schumm, 1968).

hillslope and collect in valleys until stream beds are aggraded to the point where a major storm or series of wet seasons mobilizes them, and delivers them to the trunk stream. Thus, even when a drainage basin is at maximum extension, and under conditions of stable climate, sediment does not move from the basin until trunk streams and major tributaries are aggraded to states of metastability. At that point, a threshold event may trigger an erosional episode that delivers a large amount of sediment to the fan (Fig. 5.26).

Periods of erosion would produce essentially equally-spaced pulses of coarse sediment, whereas lower yields of finer-grained sediment would be supplied during intervening periods. The process would result in the formation of fining-upward cycles on the fan and in the basin plain.

Numerous workers have alluded to the effects that the absence of vegetation must have had on runoff and sediment yield, but for many alluvial fans, these effects may have extended beyond the Devonian. During early stages of plant

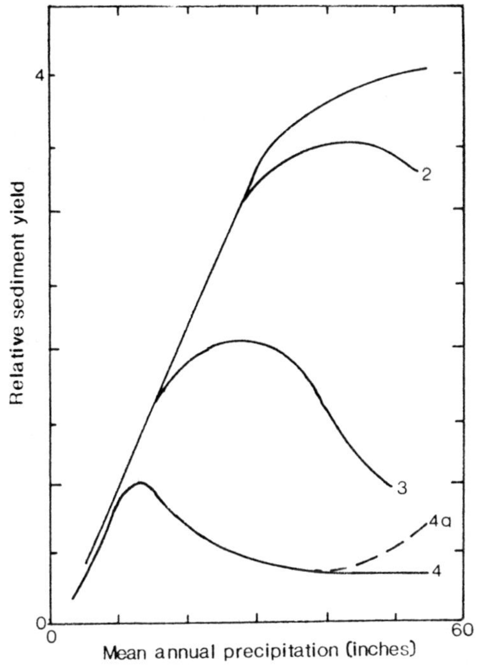

Figure 5.25 Hypothetical series of curves showing the relations between precipitation and sediment yield relative to present day (1) prevegetation; (2) to Carboniferous; (3) Carboniferous to early Tertiary; (4) post-early Tertiary (after Schumm, 1968)

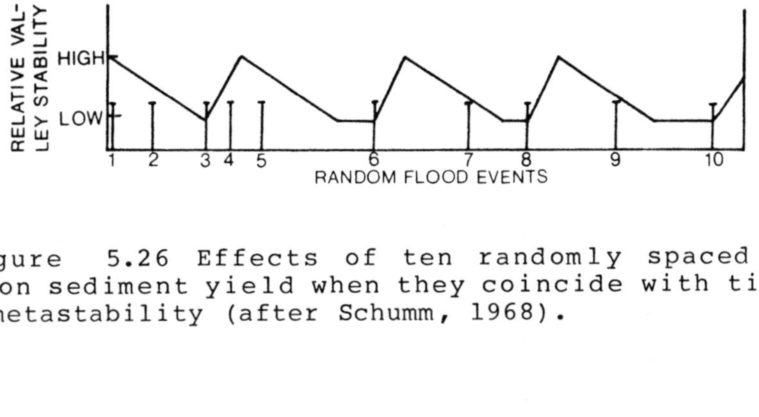

Figure 5.26 Effects of ten randomly spaced flood events on sediment yield when they coincide with times of valley metastability (after Schumm, 1968).

colonization, vegetation was probably confined to coastal plains and humid region valleys, and it is likely that interfluves in highland terrains experienced climatic conditions that were too harsh for primitive vegetation to survive on them (Schumm, 1977). Vegetation was probably present as early as the Silurian on coastal alluvial fans that stabilized their surfaces and acted as baffles to trapped fine grained sediments on their inactive portions. The response of upland source areas to changes in climate, however, probably remained unchanged from prevegetation times.

During early stages of plant colonization, runoff would have increased directly with increasing precipitation from arid to semi-arid conditions, but it would have declined as humid conditions were attained and primitive plants were able to survive and reduce runoff through transpiration and interception (Fig. 5.24). Sediment yield would have increased with increasing precipitation to the point where climatic conditions were sufficiently humid to support vegetation as galleria forests in source area valleys. At this point, the effects of reduced runoff and the binding action of plants would have slowed the rate of increase, and sediment yield would actually have declined under very humid conditions (Fig. 5.25).

The colonization of land surfaces in semi-arid regions during the Carboniferous by hardier plants significantly altered runoff characteristics and sediment yield because they promoted the retention of precipitation as soil moisture and they stabilized slopes. After the Carboniferous, runoff increased with increasing precipitation to the point where semi-arid conditions were attained. A vegetative cover was established on land surfaces that acted to slow the rate of increase temporarily (Fig. 5.24). The sediment yield would have approximated that of present-day conditions where yields decline as semi-arid conditions are attained (Fig. 5.25). It would not be until the Miocene, however, when grasses had evolved and vegetative covers could be established on upland interfluves under semi-arid conditions, that present-day conditions were established (Schumm, 1977) (Fig. 5.25).

EXTRINSIC CONTROLS ON ANCIENT FAN SEQUENCES

Devonian of Western Norway

Among the most intensely studied ancient fan sequences are those in the Devonian System conglomerates of western Norway. There, breccias, conglomerates, and sandstones were deposited on alluvial fans prograded into grabens or half-grabens (Nilsen, 1969; Steel, 1976; Steel, *et al*, 1977). A humid climate is suggested by abundant floral remains, including *psilophyton*, in the Solund Basin, and by the presence of widespread, non-evaporitic lacustrine sediments in both the Solund Basin and the Hornelen Basin.

Breccias in the Solund Basin consist of angular fragments in a matrix of feldspar-rich sand and clay. They accumulated on a highly irregular surface developed on basement rock and they probably represent talus deposits and debris flow deposits very near the fault scarp (Nilsen, 1969) (Fig. 5.27).

Breccias interfinger with and are overlain by conglomerate. The conglomerate consists of clasts to boulder size in thick, massive beds separated by thin discontinuous beds of sand in planar laminae or very low-angle cross-beds. The conglomerates probably were deposited on longitudinal bars or as diffuse gravel sheets that were set in motion during flood periods. The thin sandstone lenses probably represent deposition in lows on bar surfaces during waning flow (Nilsen, 1969).

Conglomerates are overlain by sandstones in trough and tabular cross-beds or horizontal planar laminae. Beds of conglomerate are common near the base and thin lag deposits occur elsewhere in the section. Gravel lenses probably accumulated on longitudinal bars whereas cross-bedded sandstones were deposited as transverse bars or in interbar channels.

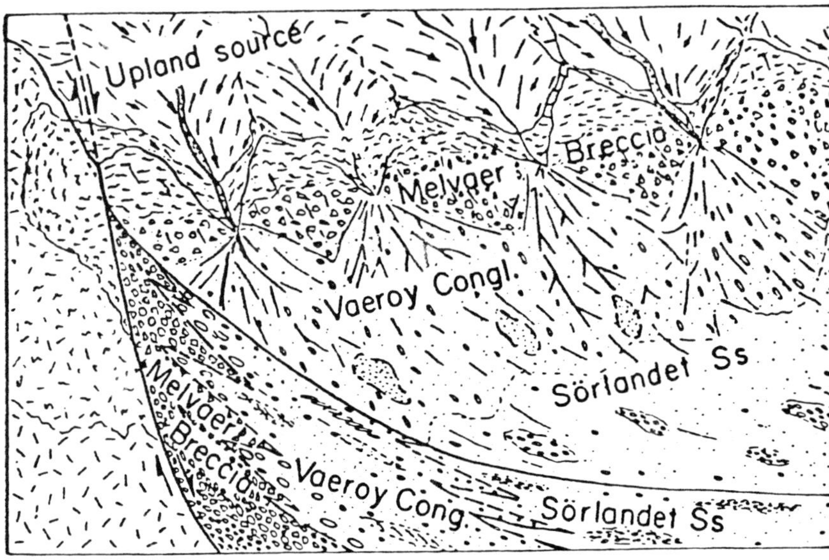

Figure 5.27 Diagrammatic representation of the facies distribution in the Solund Basin of western Norway (after Nilson, 1969).

The relative thickness of the conglomerate suggests that uplift was continuous in the source area. Nilsen (1969) ascribes the gradual overlap of the conglomerate by the sandstone facies (Fig. 5.27) as a response to erosion of the source area and retreat of the fault scarp. It is equally likely, however, that fan retrogradation occurred in response to evolution of the drainage system in the source area. Erosion of the interfluves reduced the quantity and caliber of the sediment being delivered to the streams, and widening of valleys increased the storage capacity of the flood plains, especially for coarse fractions.

Alluvial fan sediments also occur along the margin of the Hornelen Basin to the north where they display a marked cyclicity. Coarsening-upward cycles, up to 100 m thick, are in turn, are composed of coarsening-upward subcycles 10 to 25m thick (Steel, 1976, Steel, *et al*, 1977). Fan sediments were deposited by both debris-flow and stream-flow processes, and the cycles are, in turn, grouped together into sequences thousands of meters thick consisting of sediments deposited primarily by one or the other of these mechanisms (Fig. 5.28).

The fine-grained units at the base of the cycles consists most commonly of sandstones that were probably deposited on alluvial plains of the basin floor. Fan progradation into the basin produced coarsening-upward cycles by a decrease in the percentage of sandstone beds and an increase in the clast size of the conglomerates. Each cycle has a sharp, and apparently planar upper boundary where coarse fanglomerates are abruptly succeeded by sandstones of the next overlying cycle (Fig. 5.29).

All three scales of cycles probably represent responses to changes in extrinsic factors affecting the basin. The thickest sequences may represent a response to climatic change. Sequences consisting predominantly of debris-flow facies probably were deposited during relatively dry periods, whereas those deposited by stream flow presumably accumulated under wetter conditions.

The two smaller-scale subcycles, however, are attributed by Steel (1976) and Steel *et al* (1977) to subsidence of the basin along fault scarps at the margin. Cycles 10 to 25m thick formed in response to discrete episodes of faulting and the 100m-thick cycles accumulated over extended periods of frequent movement. Intermediate cycles began with a severe tectonic event that resulted in a rapid migration of basinal facies over fan facies. As equilibrium profiles were gradually reestablished, fan facies migrated more slowly basinward creating asymmetric, coarsening-upward cycles. Fan growth was apparently interrupted periodically by tectonic events of lesser magnitude that resulted in a less radical shift of facies

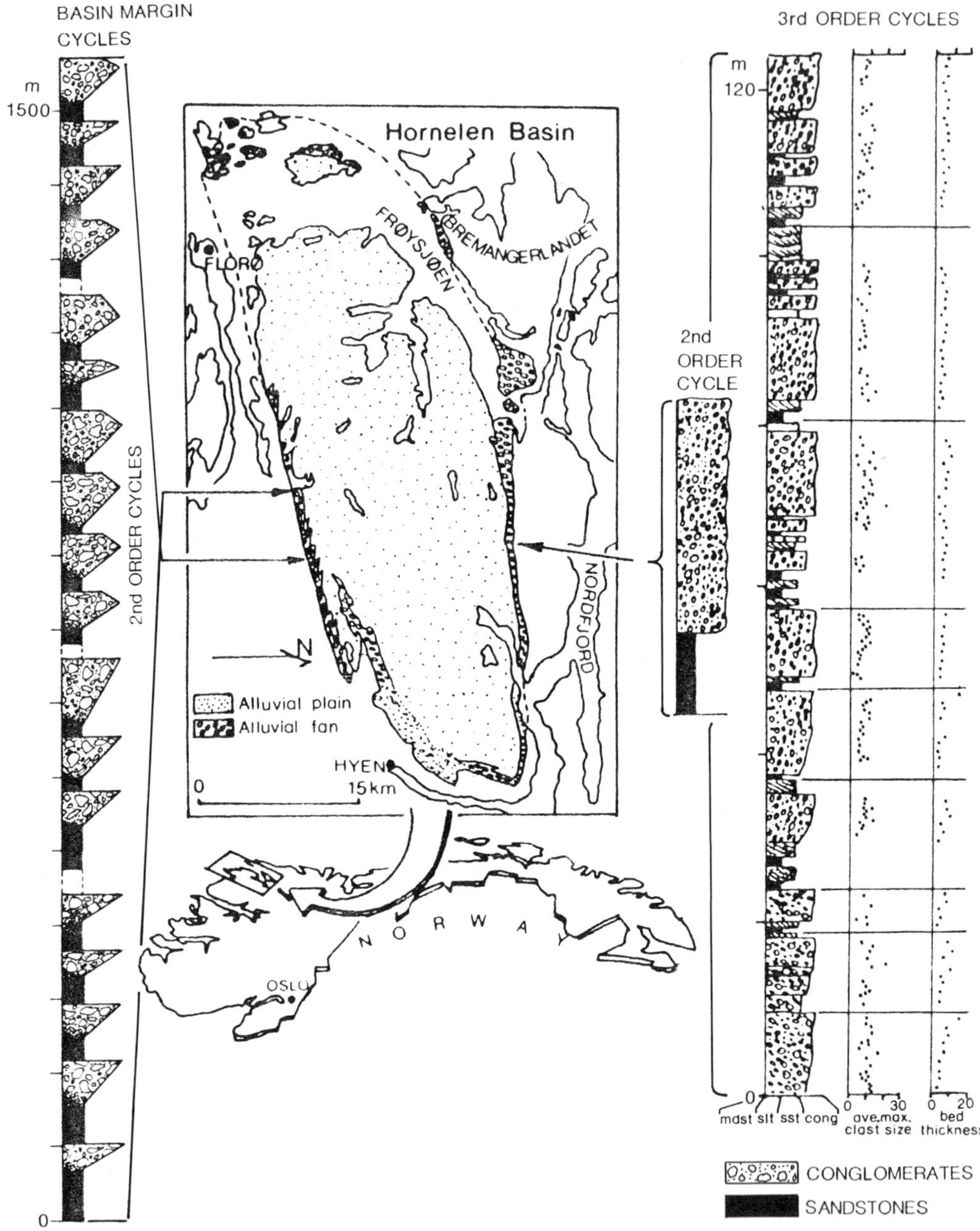

Figure 5.28 Stacked alluvial fan sequences in the Hornelen Basin of western Norway. Fan sequences on both northern and southern margins tend to coarsen upward (after Steel, 1976 and Steel, et al, 1977).

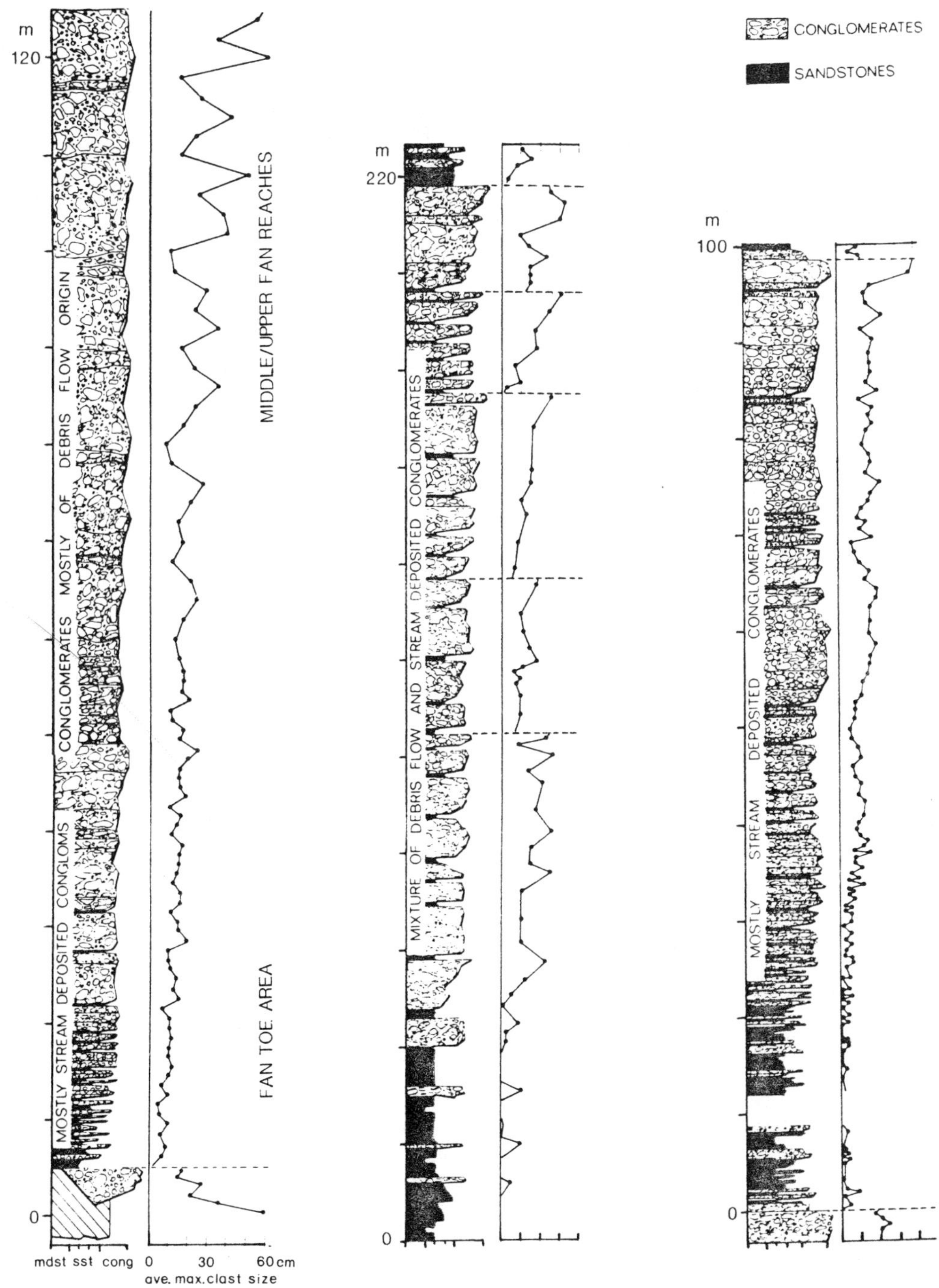

Figure 5.29 Details of coarsening and thickening-upward cycles in fanglomerates of the Hornelen Basin (after Steel, et al, 1977).

and the creation of the minor coarsening-upward cycles.

Steel, et al (1977), reject climatic variations as a viable cycle-causing mechanism because both intermediate and minor cycles occur in megasequences dominated by both debris-flow and streamflow deposits. They likewise reject an intrinsic control, such as stream avulsion, because the cyclicity can be traced not only across fan bodies, but into the basin as well.

It might be argued that the minor cycles represent the response of the drainage system to changes in base level modelled by Parker (1976, 1977). Steel, et al (1977), however, noted that subcycles thicken upward, and rejuvenation of progressively higher-order tributaries in the source area triggered by a drop in base level of the trunk stream would most likely produce subcycles that become thinner upward.

Steel (1976) also noted that fans on the northern margin of the basin rarely exceed 1 km in radius whereas those on the southern margin are commonly more than 5 km wide. In addition, fan cycles on northern fans normally are twice as thick as those in the south and the rate of downfan change of maximum clast size is greater in the north than in the south (Fig. 5.30). Therefore, within a basin, apparently subject to the same tectonic regime, the response to subsidence was the production of small steep fans along the northern margin at the same time wider fans with gentle gradients were produced in the southern margin.

Steel (1976) believes that the difference between fans on northern and southern margins may be explained by the more rapid subsidence rates on the northern side of the basin. This would account for the thicker sequences seen on the northern fans, and could account for differences in fan radius as well. Source area drainage basins on the northern margin would have been constantly subjected to rejuvenation. Downcutting would have predominated over lateral extension resulting in the development of smaller drainage basins that could only have supported fans of limited size. The southern margin, however, experienced less subsidence so that larger drainage basins, capable of supplying larger fans, could have developed (Fig. 5.31). A possible modern counterpart for this situation may be found in present-day Death Valley where fans associated with the dip-slope of upfaulted blocks tend to be larger than those adjacent to the fault scarp margins (Denny, 1965).

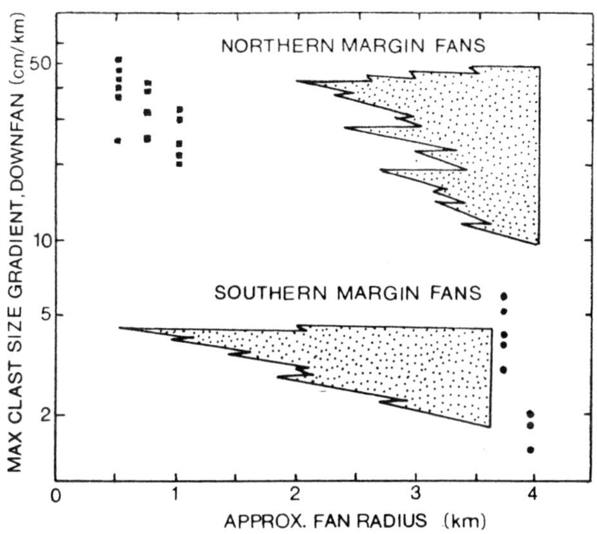

Figure 5.30 Comparison of fan shapes and clast-size variations between fans on the northern and southern margins of the Hornelen Basin (from Steel, 1976).

On the other hand, the shape of the fans may also be partly controlled by the configuration of the margin on either side of the basin much the same way that fan shape is controlled in the Ridge Basin. The small fans in the Hornelen Basin occur along the steeply dipping Nordfjord Fault and larger fans are present on the southern margin where normal faults step down to the west into the basin (Fig. 5.31).

Coarsening-upward sequences up to 100 m thick are the dominant depositional motif in both the Solund and Hornelen

Basins, but the sequences in the Solund Basin are stacked into first-order cycles, up to several thousand meters thick, that fine upward (Fig. 5.32). In the Hornelen Basin, however, no pronounced grain-size trend is present within the megasequences (Steel, 1976).

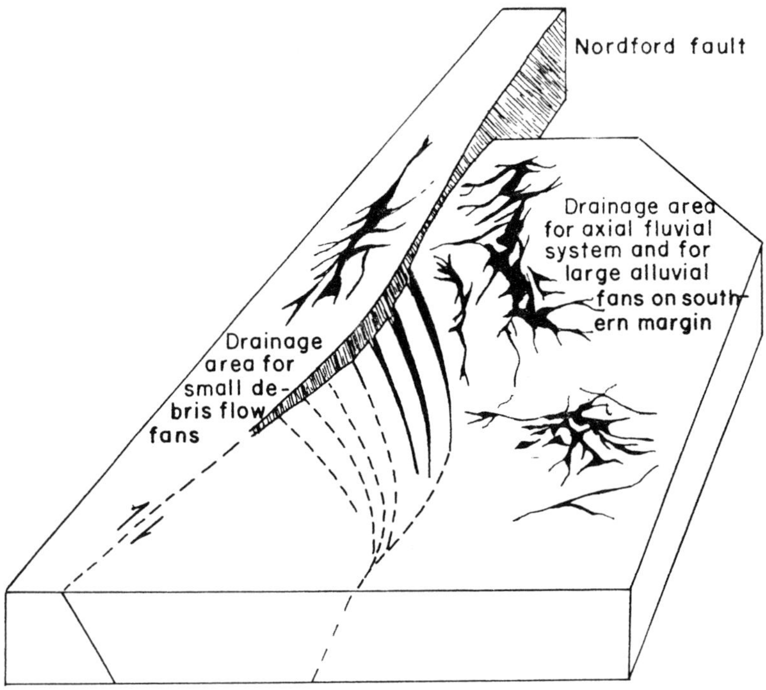

Figure 5.31 Model showing tectonic style of the Hornelen Basin and characteristics of the drainage basins on the northern and southern margins of the basin (after Steel, 1976). Compare to Figure 5.17.

The two basins also differ with respect to the dominant basin fill. Alluvial-fan conglomerates are widespread throughout the Solund Basin suggesting that fans there were considerably larger than fans in the Hornelen Basin which were restricted to the margins of a basin filled dominantly by sandy braided-stream sediments.

These differences may be explained by the two different tectonic styles responsible for the basin. Normal faulting at the basin margin continuously created high relief in the uplands bordering the Solund Basin (Steel, 1976), and provided a sufficient source for the large volume of conglomerate in the basin. Fining-upward basin-filling sequences were deposited in response to progressive lowering of relief in the source area and minor coarsening-upward sequences were deposited as streams in the source area were

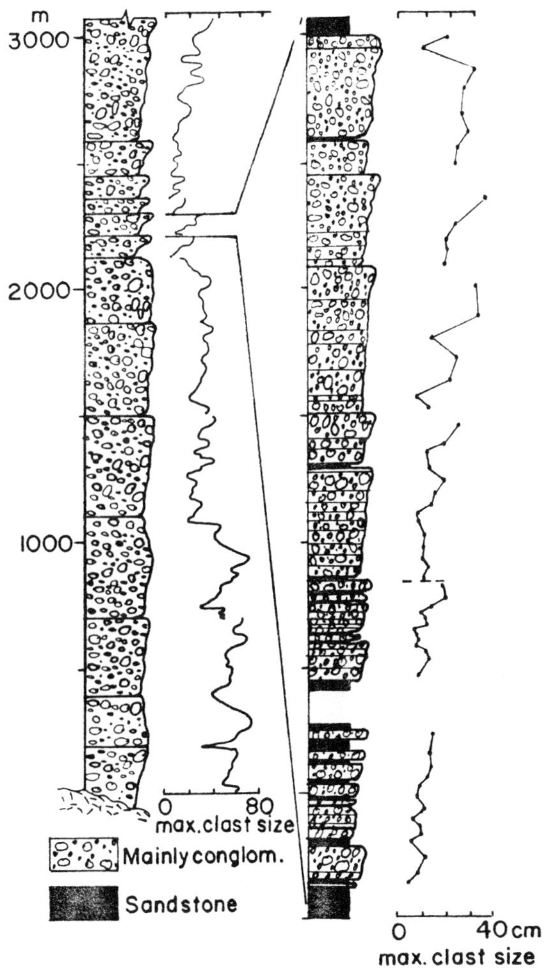

Figure 5.32 Stacked coarsening-upward sequences superimposed on a fining-upward basin-filling sequence in the Solund Basin (after Steel, 1976).

periodically rejuvenated. The thin coarsening upward sequences that are superimposed on these minor cycles may represent complex response of source area streams to lowering of base level (Fig. 5.29).

Steel (1976) suggests that the fining-upward basin-filling cycles may have formed in response to backfaulting at the margins of the basin and that the minor cycles represent response to individual tectonic episodes. Geometric considerations, however, suggest that this was not the case. Lateral displacement of fault-planes must be significant in order to induce appreciable facies migration in the basin. However, each episode of backfaulting would effectively enlarge the basin, and the Solund Basin would have to be considerably larger than it actually is in order to accommodate the numerous minor cycles present in the basin fill.

The Hornelen Basin formed as a consequence of dilation and subsidence related to wrench faulting (Fig. 5.31). The basin grew by lateral accretion and horizontal tectonic movements were more important than vertical ones (Steel, 1976). Fan bodies along uplifted margins are relatively small, and the basin filled dominantly by streams flowing longitudinally down the basin to the west.

Dominant uplift was at the east end of the basin where the largest fans occurred. As the fault was displaced laterally, the basement was onlapped to the east and individual fan systems were created that successively overlapped each other. In effect, each new tectonic episode created a new fan system and prevented the establishment of any pronounced vertical grain-size trends.

Carboniferous of Spain

Exceptionally well-described alluvial-fan sequences also occur in the Carboniferous of Spain (Heward, 1978a, 1978b). First-order cycles in thin sequences have already been discussed in terms of the intrinsic mechanisms that operated during their deposition, but they may also be used to illustrate how fan systems respond to renewed tectonism.

Proximal facies of these fans consist predominantly of conglomerates organized into two thinning and fining upward sequences up to 150 m thick. The upper sequence is coarser-grained than the lower one and individual units within it tend to be somewhat thicker. The basal sequence rests unconformably on folded Precambrian rocks and the contact between the two sequences is marked by a pronounced and abrupt change in grain size (Fig. 5.33).

Heward (1978b) interprets these two sequences as responses to uplift followed by scarp retreat. Scarp retreat is a concept commonly used to explain the occurrence of thinning and fining-upward sequences in alluvial fan deposits, but the ability of this mechanism to produce such an effect must be questioned. Except in the case of scree deposits, fans are not fed directly from a scarp, but instead receive the bulk of their sediments from drainage basins developed in the hinterlands behind the scarp. It is more likely that these two sequences formed in response to tectonic episodes followed by progressive lowering of relief in the source area.

The character of the two sequences appears to conform reasonably well to the ideal sequences expected to develop in response to an initial phase of uplift followed by secondary phases. The initial production of sediment from the source area should be relatively fine-grained because coarse sediment is being preferentially stored in source area valleys. Any subsequent periods of uplift, however, would cause erosion of these stored sediments and result in the delivery of very coarse sediment to proximal fan areas.

The large-scale sequences may be subdivided into minor cycles 10-65 m thick. Like the larger-scale sequences, these thin and fine upward and each cycle is terminated abruptly by the basal coarse-grained sediments of the overlying cycles (Fig. 5.33). Heward (1978b) interpreted these cycles to represent gradual abandonment of fan segments, but it is possible that they formed in response to

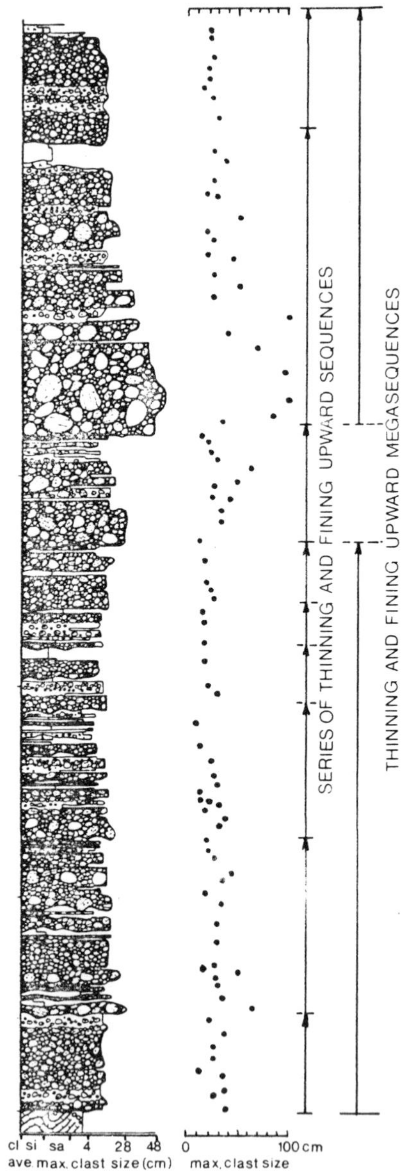

Figure 5.33 Vertical sequence in proximal facies at the Matallana coalfield (after Heward, 1978a).

intrinsic stress in the system. In the lower thick sequence, for instance, the coarse basal deposits thin and fine upward to a mid-portion where no pronounced grain size trend is evident in a stacked series of fining/thinning-upward cycles (Fig. 5.33). These cycles may represent alternating periods of fan aggradation and fan-head entrenchment that occurred on a dynamically stable fan during a period of maximum extension of the drainage net in the source area. Cycles in the upper thick sequence may represent a complex response to uplift where a series of progressively thinner-bedded and finer- grained cycles were produced by periodic entrenchment of the trunk stream following the initial period of rejuvenation (Fig. 5.33). It may be significant that the coarsest sediments in the sequence occur at the base of the upper sequence. This would be expected in the case of rejuvenation where coarse sediments, stored in source area valleys, are mobilized when valleys respond to renewed uplift by downcutting their valleys.

Mid-fan and distal-fan sequences are much more varied. In mid-fan sequences, minor cycles 5-20 m thick occur that may consist of coarsening/thickening-upward conglomerates, cross-bedded sandstones that thin and fine upward gradually, or conglomerate and sandstone sequences that thin and fine-upward more abruptly (Fig. 5.34).

Distal fan sequences consist primarily of mudstones and sandstones in sequences up to 20 m thick that thicken and coarsen upward or thin and fine-upward asymmetrically, or in symmetrical sequences that initially coarsen and thicken upward and then thin and fine-upward (Fig. 5.35).

Mid-fan and distal fan cycles are grouped into three larger-scale sequences where they can occur singly or combined and where they may display sharp or gradational contacts both vertically and laterally (Fig. 5.36). The basal of the three large-scale sequences thins and fines upward and probably represents the down-fan equivalent of the basal proximal fan sequence. The upper two, however, probably formed in response to renewed tectonism (Heward, 1978b), and their internal organization probably reflects complex response to tectonism and the operation of intrinsic mechanisms.

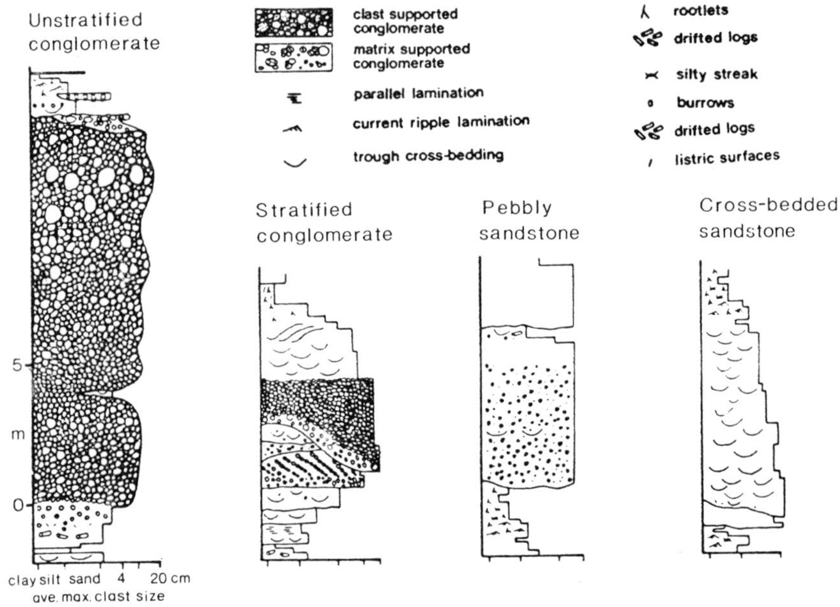

Figure 5.34 Four types of mid-fan sequences in basal fan deposits at La Magdalena, Spain (after Heward 1978a).

Coarsening-upward conglomerate cycles in the mid-fan, for instance, were deposited by prograding lobes formed during episodic rejuvenation of the trunk stream in the source area. Conglomerates and sandstones that thin and fine-upward abruptly may represent channel abandonment and cross-bedded sandstones that gradually fine upward were deposited at the distal end of lobes undergoing abandonment.

Symmetrical cycles in the distal fan probably represent a complete sequence of sediments deposited by prograding minor lobes whose formation may have been initiated by episodes of fan- head entrenchment or rejuvenation of source area streams. The coarsening/thickening-upward parts of the cycle accumulated during the progradational phase. The thinning/fining upward part was deposited during a retrogradational phase in response to a gradual decline in the rate of erosion as channel gradients approached equilibrium conditions.

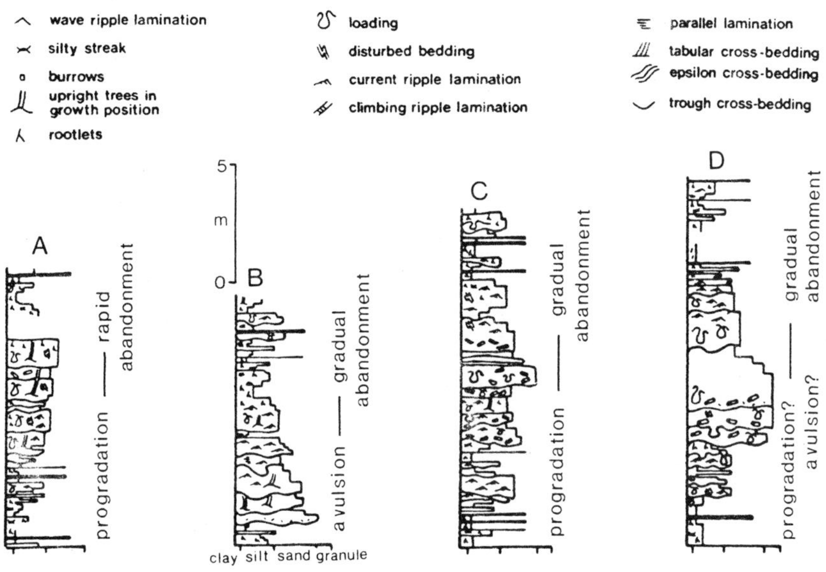

Figure 5.35 Distal fan sequences at La Magdalena, Spain showing asymmetric thickening and coarsening-upward sequence (A); asymmetric thinning and fining-upward sequence (B); symmetric thickening and coarsening then thinning and fining upward sequences (C, D) (after Heward, 1978a).

Asymmetric sequences formed in response to events that were either initiated or terminated abruptly. Fining/coarsening-upward sequences may have accumulated when channel avulsion or stream capture occurred. In contrast, thickening/coarsening-upward ones formed when lobe progradation was terminated by channel abandonment.

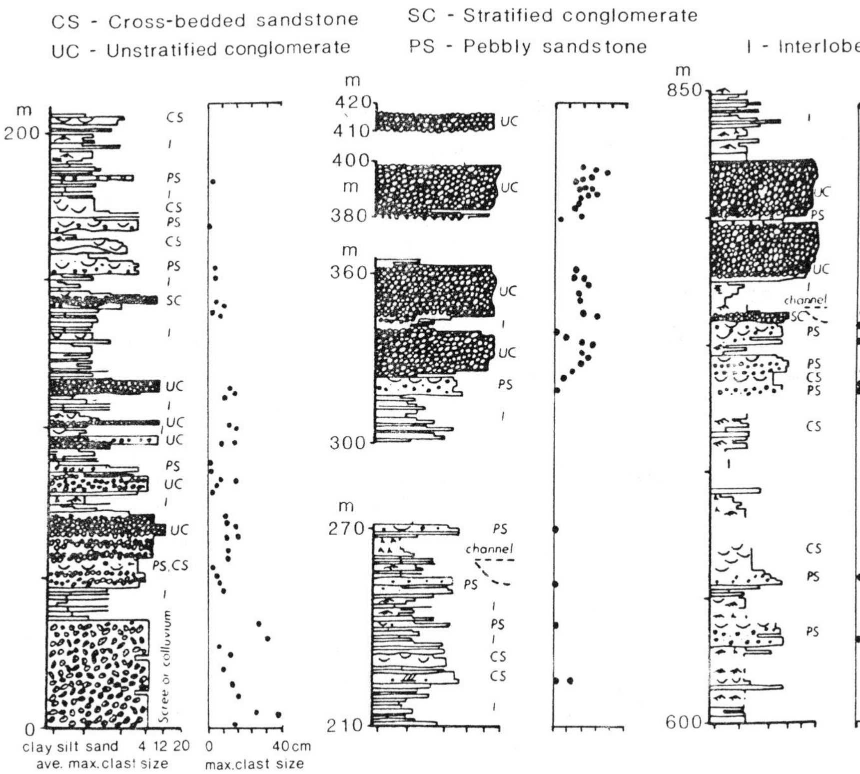

Figure 5.36 Section showing the organization of four facies into a vertical sequence typical of those found within mid-fan lobe deposits at La Magdalena, Spain (after Heward, 1978a).

6

Diagenesis of Alluvial-Fan Deposits

INTRODUCTION

Diagenesis is the most complex of the major processes which influence the origin of sedimentary rocks. This is because it is a function of a multivariate system. And variables begin affecting the ultimate diagenetic products as early as when weathering produces detritus in a source area, and do not cease affecting the diagenetic products until the initiation of the next sedimentary cycle. Nature of parent rock, relief, climate, transportational-depositional history and burial/exhumation conditions together form the multivariate system which controls the diagenetic mineral associations (Fig. 6.1). In alluvial fans and fan deltas, where each of these factors potentially can fluctuate widely, understanding and predicting diagenetic patterns in reservoirs can be exceedingly difficult. And certainly no two fans will experience the same diagenetic histories. In this chapter we will discuss each of the major variables controlling diagenesis in fans and fan deltas in a conceptual way--in a way that should permit the explorationist to think about diagenesis uniquely for a given system. We will not attempt to construct diagenetic models for alluvial fans and fan deltas. Such models would be artificial oversimplifications and would be of little value in rigorous reservoir analysis.

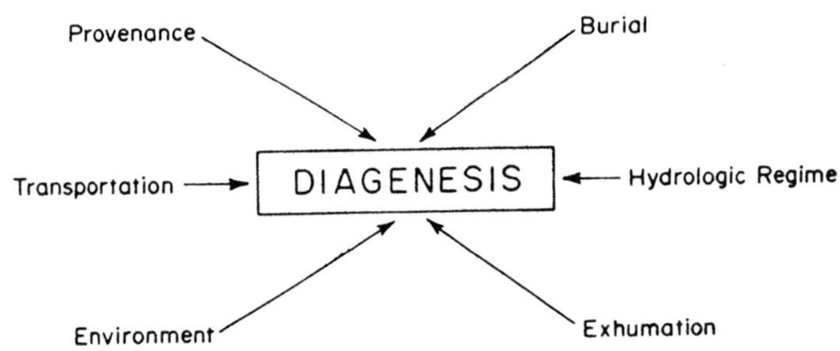

Figure 6.1 Variables controlling diagenesis in an alluvial fan, fan-delta system.

Consideration of the diagenesis of alluvial fans and fan deltas is hampered by little previous study of the diagenesis of conglomerates. Fortunately, recent interest in the sedimentology of conglomerates is catalyzing interest in their diagenesis as well. Several critical and useful recent articles on the topic include Candido and Wardlow (1985), Dutton and Land (1985), Stonecipher et al (1984), Melvin and Knight (1984), Varley (1984), Cant and Ethier (1984), and Taylor (1978). These articles, especially Candido and Wardlow (1985) and Stonecipher et al (1984), place heavy emphasis on local environmental control on diagenesis in coarse clastic facies in fan and fan deltas. Although we would tend to agree that depositional facies exerts the dominant control on diagenetic facies, we caution that it is certainly not the exclusive control.

POROSITY, PERMEABILITY AND GRAIN SIZE DISTRIBUTIONS

Although porosity and permeability obviously are affected by diagenesis, in coarse clastic facies the effects of diagenesis on the two properties should be considered separately from the effects of the grain-size distribution itself. This was first elegantly demonstrated by Clarke (1979) and later reiterated by Cant and Ethier (1984).

All other factors being equal, most notably diagenetic history, well sorted sandstones and unimodal or open-framework conglomerate (i.e. conglomerates lacking a sand matrix) have highest initial porosity, but open framework conglomerates additionally have the highest permeability (Fig. 6.2). At constant porosity, permeability is proportional to the square of the grain size for perfect spheres, thus the explanation for the difference in permeability. Smallest porosity occurs in clast-supported conglomerate where interstices are filled with sand. This is because a bimodal conglomerate can be thought of as having the porosity of the sandstone component minus the porosity of the volume of sand represented by the non-porous pebbles or cobbles.

Three possible sites exist in fans and fan deltas for optimum accumulation of open framework conglomerates possessing high initial porosity and permeability; (i) areas of sieve deposition in the proximal fan, (ii) longitudinal bars in channels where fine-sediment bypassing is prevalent, and (iii) the upper shoreface or beach where winnowing is effective. Most forms of sediment gravity flows will give rise to bimodal conglomerates characterized by low initial porosity and permeability. Distributary-mouth diamictites will possess small values because of the abrupt deposition of sediment, and resultant lack of selective sorting, at the distributary mouth.

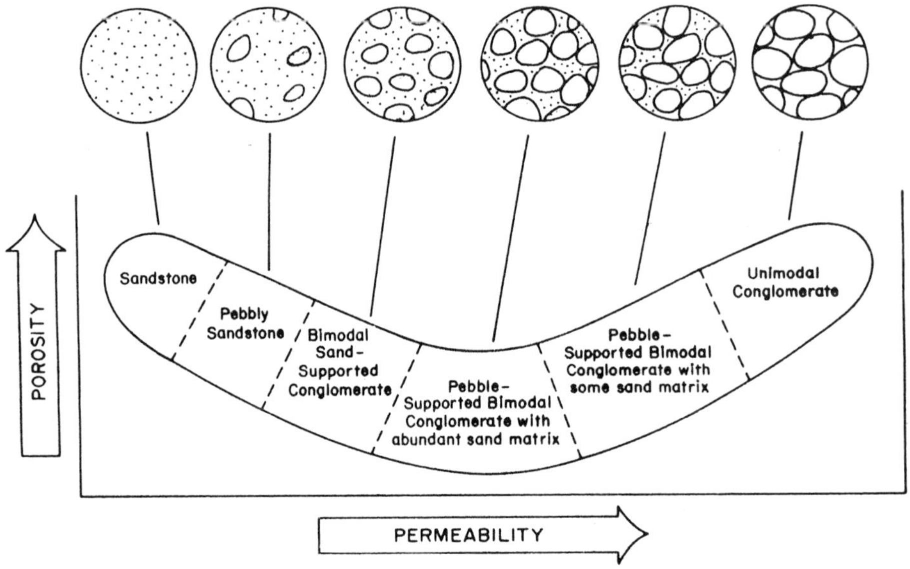

Figure 6.2 Relationship between grain size, sorting, and porosity and permeability in sandstone, conglomeratic sandstone and conglomerate (after Clarke, 1979).

VARIABLES AFFECTING DIAGENESIS

Parent Rock

Parent rock, dictated largely by tectonic setting, is the primary control on framework grain composition, and framework composition influences diagenesis in three important ways. First, unstable grains of the type derived from older volcanic rocks (e.g. rock fragments and Ca-rich feldspars) quickly alter to pore-filling clays and zeolites in diagenesis (Galloway, 1979; Surdam and Boles, 1979). Compaction of the clays during burial diagenesis further reduces porosity. But, unstable framework grains are also most susceptible to complete dissolution resulting in high secondary porosity. Traditionally, the affect of porosity reduction by clay and zeolite authigenesis, particularly in volcaniclastics, has been assumed to be greater than porosity enhancement by secondary dissolution (Galloway,

1979). However, Mathisen (1984, p. 191) suggested that this is not always the case. He reported an increase of porosity (25% to 37%) in volcaniclastic sand down to a burial depth of 900 m and attributed this to early diagenetic grain dissolution. Earlier, Walker, et al (1978) demonstrated that significant early secondary porosity can form in first-cycle desert alluvium. Certainly, the sharp hydraulic gradients of alluvial fans are ideal for promoting flux of meteoric water through alluvium. This keeps the ionic concentrations of pore waters low, which in turn stimulates grain dissolution, but retards local re-precipitation.

The third control of parent material on diagenesis involves the influence of framework composition on pore water chemistry, particularly the Ca^{2+} ionic activity of pore water, which dictates amounts of calcite cement. In some cases, calcite cement occludes pore space, but in other cases late burial dissolution of the cement can enhance secondary porosity. For example, porosity in granite wash in alluvial fan-delta facies in the Mobeetie field of the Texas panhandle was reduced where calcareous framework grains were recrystallized into pore-filling calcite cement (Dutton and Land, 1985). In contrast, in the Ivishak Formation, an alluvial-fan reservoir in the Prudhoe Bay field, early calcite cement in the form of caliche apparently dissolved during later diagenesis to improve porosity (Melvin and Knight, 1984).

Alluvial fans are prime sites of caliche development, because of their subaerial exposure and common occurrences in arid climates where Ca^{2+} activity of groundwater is high because of low fluid flux. However, often it is not realized that caliche can form under relatively moist conditions if large Ca^{2+} activity can be maintained in spite of low evaporation and groundwater flux. The simplest way to maintain the large Ca^{2+} activity is to have carbonate-rich parent rocks which give rise to abundant carbonate framework grains. Abundant windblown carbonate silt can serve the same purpose and has been demonstrated to be an important source of calcite cement in modern day fans (Lattman, 1973).

Because of the strong parent rock control on framework composition, and thus diagenesis, thick alluvial fan sequences are logical places to expect systematic vertical changes in diagenesis which may be independent of burial-induced diagenetic phenomena. For example, high-angle bounding faults of the type associated with alluvial fan formation are ideal for generating simple unroofing sequences which expose basement rocks. Within thick alluvial fan sequences which are a product of the unroofing, carbonate and detrital chert clasts and quartz derived from older sedimentary supra-crustal rocks should decrease upward in the sequences, and feldspar and crystalline rock fragments derived from deeper in the crust should increase. These changes in framework composition could exert important control on stratigraphic variations in diagenetic history.

Climate

The full importance of climate in early diagenesis of alluvial fan sediments only recently has become appreciated (Dutta and Suttner, 1986) although Walker, et al, 1978) first called attention to it. Climate exerts control indirectly on diagenesis through its influence on pore-water flux rate and thus pore water chemistry. In wet climates, the flux is great, cations such as Fe, Mg, and K are moved away rapidly from sites of dissolution of framework minerals and pore water remains undersaturated with respect to these ions. This enhances early dissolution of framework grains, particularly the feldspars, and gives rise to large ratios of kaolinite to smectite plus chlorite in the early authigenic mineral association. Just the opposite relationship exists on alluvial fans in arid climates. Dutta and Suttner (1986) have recognized significant differences in both the framework and authigenic mineral composition of first-cycle alluvial sands from contrasting climates (Figs. 6.3, 6.4). Because these differences largely are imparted during pedogenesis, they are especially sharp on alluvial fans where soil-forming processes are operative on inactive parts of fans.

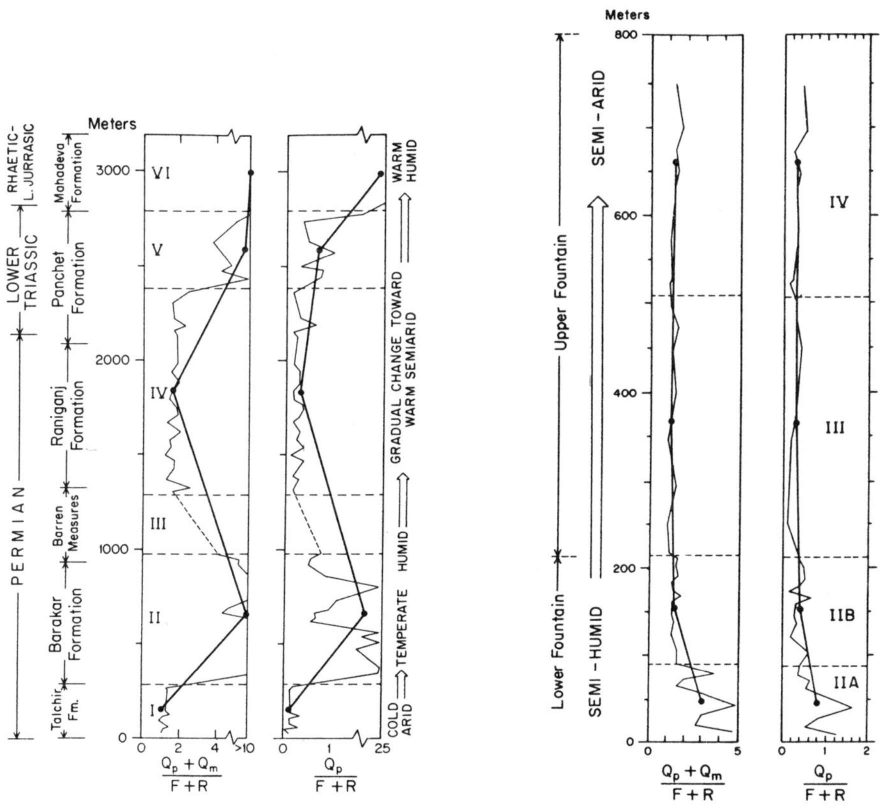

Figure 6.3 Stratigraphic variation in amount of polycrystalline quartz (Qp), monocrystalline quartz (Qm), feldspar (F) and rock fragments (R) in arkosic sandstone as a function of changing paleoclimate. Data are from sandstones in the Gondwana Supergroup from India and the Fountain Formation from near Colorado Springs (from Suttner and Dutta, 1986).

Finally, climate-controlled early diagenesis on alluvial fans, especially in arid climates where ionic activity of pore water is great, might quickly bring about generation of clay rims on framework grains. These clay rims, if well developed, inhibit overgrowth formation on quartz and feldspar which can occlude pore space.

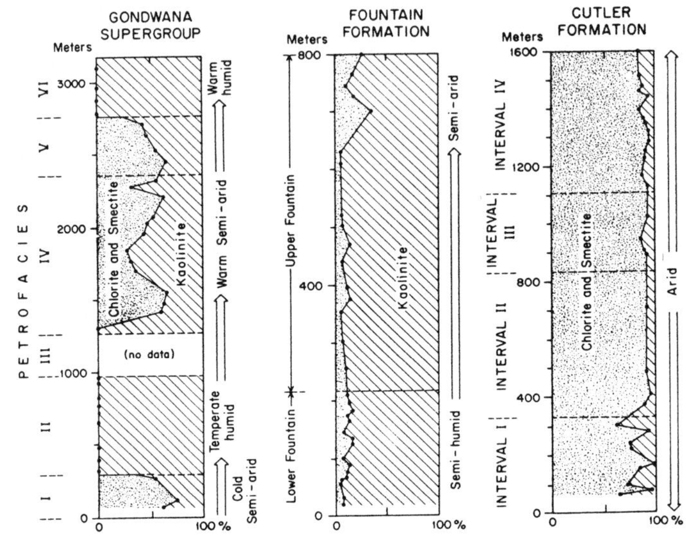

Figure 6.4 Stratigraphic variation in authigenic mineralogy of arkosic sandstone as a function of paleoclimate. Data are from the Gondwana Supergroup in India and the Cutler and Fountain Formations in Colorado. Note how the ratio of total chlorite plus smectite to kaolinite is greatest for periods of maximum aridity (from Dutta and Suttner, 1986).

Depositional Facies

Both the texture and framework composition of fan and fan-delta detritus are markedly influenced by transportational/depositional history. Therefore, it is not surprising that a close correlation exists in fans and fan deltas between diagenetic facies and depositional facies. Moreover, fans and fan deltas are characterized by closely juxtaposed environments of sharply different "energy". Thus, abrupt and magnified stratigraphic and lateral variations in diagenetic patterns should be the exception rather than the rule.

Transportation distance and environment initially affect diagenesis through their control of framework composition. The primary trend is reduction in feldspar and other labile minerals associated with transport and ultimately, in fan deltas, swash/backwash movement in high energy shoreline environments.

Suttner *et al* (1981) conclude that a 15-20% reduction in both feldspar and rock fragments is possible in the initial 75 km of river transport of first-cycle sand. In littoral environments, an additional 65% of the feldspar and close to 80 percent of the entire amount of rock fragments can be destroyed. Although the transport factor is not applicable to most fans, because of their small radius, the more compositionally mature sand in the shoreline facies of fan deltas would be less conducive to development of secondary porosity through unstable grain dissolution. Furthermore, the greater quartz content of the shoreline sandstones provides abundant nuclei for quartz overgrowth development. And the absence, due to winnowing, of protective clay coats around quartz grains leaves them ripe for overgrowth development. Basically, for the above reasons Stonecipher *et al* (1984) discovered greater porosity in the fluvial facies of fan deltas than the marine facies. A related example of environmental control on diagenesis through framework composition is Dutton and Land's (1984) observation that marine- reworked sandstone in fan deltas has lower porosity than the non-marine facies, because marine reworking brought about incorporation of calcareous fossil fragments into the sand. These calcareous grains recrystallized to form pore-filling calcite cement. Conceivably, in some cases the formation of early calcite cement of this origin might actually improve porosity, if, as is likely to be the case, it undergoes later dissolution.

The key textural control on diagenesis influenced by environment is mud content of the sand, and, as discussed earlier, sand-matrix content of conglomerate. Presence or absence of mud can play a complicated role in diagenesis. The presence of some mud is good because it retards overgrowth development; too much mud, however, upon compaction clogs pore throats. Thus, lower shoreface and distributary mouth sands generally have lower porosity and permeability. Early clogging of pore throats by clay also inhibits flow of pore fluids which cause grain dissolution and formation of secondary porosity. Moreover, Stanton (1977) has shown that small amounts of detrital mud intermixed with calcite cement actually reduce permeability enough to inhibit flow of formation water, if not to detour the water through more permeable facies. Bioturbation often mixes detrital mud into sand. Lower shoreface and backshore sands in fan deltas tend to be burrowed and, therefore, have reduced porosity and permeability (Tilley and Longstaffe (1984).

Although differences in environmental control on diagenetic facies will be less pronounced on the subaerial part of fan-deltas, nevertheless they do exist. Pedogenic processes will dominate on inactive parts of fans to produce early-formed mud matrix. Levees and bar-tops in channels will contain more detrital mud. And fine-grained gravity flows (e.g. mud flows) can be major permeability-porosity barriers. An excellent discussion of the multi-scale heterogeneity of porosity and permeability within oil and gas reservoirs in subaerial alluvial-fan facies, and the effect of this heterogeneity on tertiary recovery processes is found in Candido and Wardlaw (1985).

Hydrologic Regime

The hydrologic regime of an alluvial fan is relatively simple, that of a fan delta more complex especially if it is associated with a relatively deep sedimentary basin (Fig. 6.5). Alluvial fans largely undergo cementation by flux of meteoric water. The greater the flux, the greater the secondary dissolution. Fan-delta facies are more susceptible to influences of water derived from compaction and expulsion of pore waters from basin-axis sediments, as discussed above, and thermobaric waters released by dehydration reactions of clays at high temperature and pressure in deeply buried argillaceous rocks. The latter is a "good news-bad news" situation in that organic acids derived under deep pressure promote dissolution of calcite and other framework minerals. But the same fluids which transport the acids carry the ions which precipitate pore filling clays in sandstone. For this reason potential reservoir rocks in fan deltas which are encased by, or connect downdip with argillaceous rocks may suffer much loss of initial porosity through late-stage clay authigenesis.

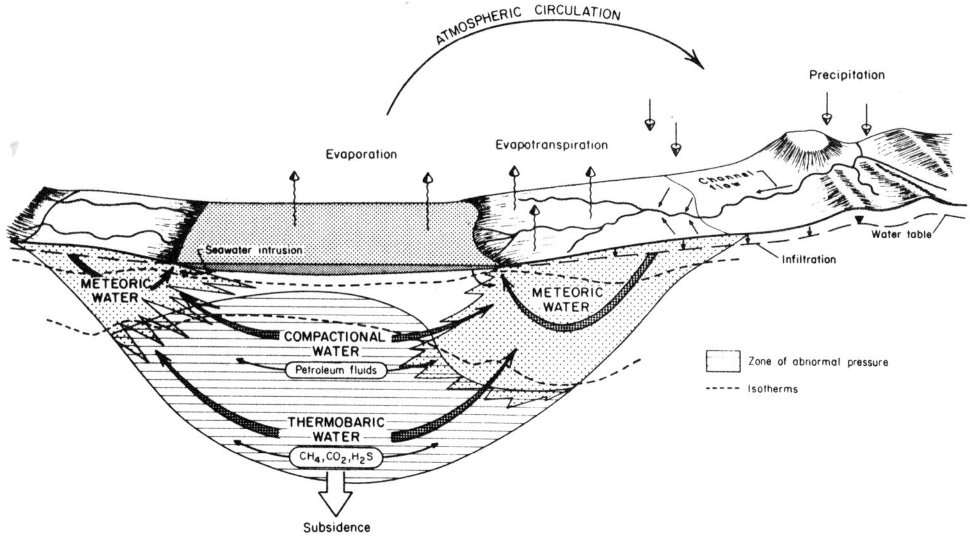

Figure 6.5 The hydrologic regimes within a major sedimentary basin. Alluvial-fan diagenesis largely is controlled by the meteoric regime. Fan-delta facies can be affected during diagenesis by compactional, thermobaric and meteoric waters (from Galloway and Hobday, 1983).

References

Ackroyd, Peter, and Blakely, R. J., 1984, En masse debris transport in a mountain stream: Earth surf. Proc. Landforms, v. 9, p. 307-320.

Anderson, D. W., and M. D. Picard, 1974, Evolution of synorogenic clastic deposits in the intermontane Uinta Basin of Utah: in Dickinson, W. R., ed., Tectonics and sedimentation, S.E.P.M. Spec. Pub. 22, p. 167-189.

Anstey, R. L., 1965, Physical characteristics of alluvial fans: U.S. Army Natick Lab. Tech. Rept. ES-20, 109 p.

Armstrong, R. L., 1968, Sevier orogenic belt in Nevada and Utah: Geol. Soc. America Bull., v. 79, p. 429-458.

Belt, E. S., 1968, Carboniferous continental sedimentation, Atlantic Provinces, Canada: in, Klein, G. deV., ed., Late Paleozoic and Mesozoic continental sedimentation, northeastern North America, Geol. Soc. America Spec. Paper 106, p. 127-176.

Blake, M. C., Jr., et al, 1978, Neogene basin formation in relation to plate-tectonic evolution of San Andreas fault system, California, Am. Assoc. Petroleum Geol. Bull., v.62, p. 344-372.

Blissenback, E., 1954, Geology of alluvial fans in semi-arid regions: Geol. Soc. America Bull. v. 65, p. 175-190.

Bluck, B J., 1967, Deposition of some Upper Old Red Sandstone conglomerates in the Clyde area: a study in the significance of bedding: Scottish Jour. Geology, v. 3, p. 139-167.

Boothroyd, J. C., and Ashley, G. M., 1975, Process, bar morphology and sedimentary structures on braided outwash fans, northeastern Gulf of Alaska, in Jopling, A. V., and McDonald, B. C., eds., Glacio-fluvial and glacio-lacustrine sedimentation: S. E. P. M. Spec. Pub. 23, p. 193-222.

Boothroyd, J. C., and Nummedal, Dag, 1978, Proglacial braided outwash: a model for humid alluvial fan deposits, in A. D. Miall, ed., Fluvial sedimentology: Can. Soc. Petroleum Geol., Mem. 5, p. 651-668.

Boyce, R. C., 1975, Sediment routing with sediment delivery ratios: Agricultural Res. Serv. ARS-S-40, p. 61-65.

Brown, L. F., Jr., 1979, Deltaic sandstone facies of the mid-continent: in Hyne, N. J., ed., Pennsylvanian sandstones of the mid-continent: Tulsa Geological Society Spec. Pub. 1, p. 35-63.

Bryhni, I., and F. J. Skjerlie, 1975, Syndepositional tectonism in the Kvamshesten district (Old Red Sandstone), western Norway: Geol. Mag., v. 112, p. 593-600.

Bull, W. B., 1968, Alluvial fans: Jour. Geol. Education, v. 16, p. 101-106.

Bull, W. B., 1977, The alluvial-fan environment: Prog. Physical Geog. v. 1, p. 222-270.

Burgisser, H. M., 1984, A unique mass flow marker bed in a Miocene streamflow molasse sequence. Switzerland, in Koster, E. H., and Steel, R. J. eds., Sedimentology of gravels and conglomerates, Can. Soc. Petroleum Geol., Mem. 10, p. 147-164.

Candido, A., and Wardlaw, N. C., 1985, Reservoir geology of the Carmopolis oil field, Brazil: Bull. Can. Petroleum Geol., p. 379-395.

Cant, O. J., and Ethier, V. G., 1984, Lithology-dependent diagenetic control of reservoir properties of conglomerates, Falher Member, Elmworth Field, Alberta: Am. Assoc. Petroleum Geol. Bull. v. 68, p. 1044-1054.

Clarke, R. H., 1979, Reservoir properties of conglomerates and conglomeratic sandstones: Amer. Assoc. Pet. Geol., v. 63, p. 799-809.

Clifton, H. E., 1981, Progradational sequences in Miocene shoreline deposits, southeastern Caliente Range, California: Jour. Sed. Petrology, v. 51, p. 165-184.

Coleman, J. M., and Prior, D. B., 1984, Deltaic environments of deposition, in Scholle, P. A., and Spearing, Darwin eds., Sandstone depositional environments, Am. Assoc. Petroleum Geol. Mem. 31, p. 139-178.

Crowell, J. C., 1974, Origin of late Cenozoic basins in northern California: in Dickinson, W. R., ed., Tectonics and sedimentation, S.E.P.M. Spec. Pub. 22, p. 190-204.

Crowell, J. C., 1982a, The Violin Breccia, Ridge Basin, southern California: in Crowell, J. C., and M. H. Link, eds., Geologic history of the Ridge Basin, southern California, Pacific Sec., S.E.P.M., Los Angeles, p. 89-98.

Crowell, J. C., 1982b, The tectonics of Ridge Basin, southern California: in Crowell, J. C., and M. H. Link, eds., Geologic history of the Ridge Basin, southern California, Pacific Sec. S.E.P.M., Los Angeles, p. 25-42.

Crowell, . C., and M. H. Link, eds., 1982, The Geologic History of the Ridge Basin, southern California: Pacific Sec. S.E.P.M., Los Angeles, 304 p.

de Feyter, A. J., and Molenaar, N., 1984, Messiniau fanglomerates: the Colambacci Formation in the Dietrarubbia Basin, Italy: Jour. Sed. Petrology, v. 54, p. 746-758.

Denny, C. S., 1965, Alluvial fans in the Death Valley region, California and Nevada: U. S. Geol. Survey. Prof. Paper 466, 62 p.

Denny, C. S., 1967, Fans and pediments: Am. Jour. Science, v. 265, p. 81-105.

Dickinson, W. R., 1974, Plate tectonics and sedimentation in Dickinson, W. R., ed., Tectonics and sedimentation, S. E. P. M. Spec. Pub. 22, p. 1-27.

Dickinson, W. R., 1976, Plate tectonic evolution of sedimentary basins, in Dickinson, W. R., and Yarborough, Hunter, eds., Plate tectonics and hydrocarbon accumulation: Am. Assoc. Petroleum Geol. Cont. Education Short Course Note Ser. No. 1, p. 1-56.

Dunbar, C. O., 1969, Historical geology: New York, Wiley, 500 p.

Dutta, P. K., and Suttner, L. J., 1986, Alluvial sandstone composition and paleoclimate; Part II, authigenic mineralogy; Jour. Sed. Petrology, v. 56 (in press).

Dutton, S. P., and Land, L. S., 1985, Meteoric burial diagenesis of Pennsylvanian arkosic sandstones, southwestern Anadarko Basin, Texas: Amer. Assoc. Petroleum Geol. Bull., v. 69, p. 22-38.

Eckis, R., 1928, Alluvial fans in the Cucamonga district, Southern California: Jour. Geology, v. 36, p. 111-141.

Ethridge, F. G., and Wescott, W. A., 1984, Tectonic setting, recognition and hydrocarbon reservoir potential of fan-delta deposits: in Koster, E. H., and Steel, R., eds., Sedimentology of gravels and conglomerates, Can. Soc. Petroleum Geol. Mem. 10 p. 217-236.

Flint, S., 1985, Alluvial fan and playa sedimentation in an Andean arid closed basin: the Placencia Group, Antofagasta Province, Chile: Jour. Geol. Soc. London, v. 142, p. 533-546.

Flint, S., Clemmey, H., and Turner, P., 1986, The Lower Cretaceous Way Group of northern Chile: an alluvial fan-fan delta complex: Sedimentary Geol., v. 46, p. 1-22.

Fraser, G. S., and Cobb, J. C., 1982, Late Wisconsinan proglacial sedimentation along the West Chicago Moraine in northeastern Illinois: Jour. Sed. Petrology, v. 52, p. 473-491.

Fraser, G. S., Bleuer, N. K., and Smith, N. D., 1983, History of Pleistocene alluviation of the Middle and Upper Wabash Valley: in Shaver, R. S., and Sunderman, J. A., eds., Field trips in midwestern geology, Ind. Geol. Survey Field Trip Guidebook, p. 197-224.

Fraser, G. S., and Fishbaugh, D. A., 1985, Sedimentology of the Coal Creek Fan, Fountain County, Indiana: Indiana Geol. Survey Occ. Paper 48, 7 p.

Frost, E. F., Jr., Allen, T., and Fertl, W. H., 1982, Formation evaluation in granite-wash reservoirs: World Oil, v. 195, p. 121-132.

Gager, B. R., 1984, The Tiger Formation: Paleogene fluvial sediments deposited adjacent to deforming cordilleran metamorphic core complex, northeastern Washington: Sedimentary Geol., v. 38, p. 393-420.

Galloway, W. E., 1979, Diagenetic control of reservoir quality in arc-derived sandstones: implications for petroleum exploration *in* Scholle, P. A., and Schluger, P. R., eds., Aspects of diagenesis: S. E. P. M. Spec. Pub. 26, p. 251-262.

Galloway, W. E., and Hobday, D. K., 1983, Terrigenous clastic depositional systems: New York, Springer Verlag, 423 p.

Garner, H. F., 1959, Stratigraphic-sedimentary significance of contemporary climate and relief on four regions of the Andes Mountains: Geol. Soc. America Bull., v. 70, p. 1327-1368.

Glock, W. S., 1932, Available relief as a factor of control in the profile of a landform: Jour. Geology, v. 40, p. 74-83.

Gloppen, T. G., and Steel, R. J., 1981, The deposits, internal structure and geometry of six alluvial fan-fan delta bodies (Devonian-Norway) - a study in the significance of bedding sequence in conglomerates, *in* Ethridge, F. G., and Flores, R. M., eds., Recent and ancient nonmarine depositional environments: Models for exploration: S. E. P. M. Spec. Pub. 31, p. 46-69.

Gole, C. V., and Chitale, S. V., 1966, Inland delta building activity of the Kosi River: Am. Soc. Civil Engineers, Jour. of the Hydraulic Division, v. 92 (Hy2), p. 111-126.

Grover, J. A., 1984, Petrology, depositional environments and structural development of the Mineta Formation, Teron Basin, Cochise County, Arizona: Sedimentary Geol., v. 38, p. 87-106.

Hadley, R. F., and Schumm, S. A., 1961, Sediment sources and drainage basin characteristics in upper Cheyenne River basin: U. S. Geol. Survey Water-Supply Paper 1531-B, p. 137-196.

Harms, J. C. *et al*, 1975, Depositional environments as interpreted from primary sedimentary structures and stratification sequences: S.E.P.M. Short Course No. 2, 161 p.

Harms, J. C., Southard, J. B., and Walker, R. G., 1982, Structures and sequences in clastic Rocks: S.E.P.M. Short Course No. 9, 249 p.

Hein, F. J., 1984, Deep-sea and fluvial braided channel conglomerates: a comparison of two case studies, *in* Koster, E. H., and Steel, R.J., eds., Sedimentology of gravels and conglomerates, Can. Soc. Petroleum Geol. Mem. 10, p. 33-50.

Heward, A. P., 1978a, Alluvial fan and lacustrine sediments from the Stephanian A and B (La Magdalena, Cinera-Matallana and Sabero) coalfields, northern Spain: Sedimentology, v. 25, p. 451-488.

Heward, A. P., 1978b, Alluvial fan sequence and megasequence models: with examples from the westphalian D-Stephanian B coalfields, northern Spain, in Miall, A. D., ed., Fluvial sedimentology: Can. Soc. Pet. Geol., Mem. 5, p. 669-702.

Hoffman, Paul , Dewey, J. F., and Bunke, Kevin 1974, Aulacogens and their genetic relation to geosynclines, with a Proterozoic example from Great Slave Lake, Canada: S. E. P. M., Spec. Pub. 19, p. 38-55.

Holmes, A., 1965, Principles of Physical Geology, 2nd Edition: New York, The Ronald Press Co., 1288 p.

Hooke, R. LB., 1967, Processes on arid-region alluvial fans: Jour. Geology, v. 75, p. 438-460.

Hooke, R. LB., 1968, Steady-state relationships on arid-region alluvial fans in closed basins: Am. Jour. Science, v. 266, p. 609-629.

Hunt, C. B., and Mabey, D. R., 1966, Stratigraphy and structure, Death Valley, California: U. S. Geol. Survey Prof. Paper 494-A, 162 p.

Karcz, I., 1972, Sedimentary structures formed by flash floods in Southern Israel: Sedimentary Geol., v. 7, p. 161-182.

Kessler, L. G., and Moorhouse, K., 1984, Depositional processes and fluid mechanics of Upper Jurassic conglomerate accumulations, British North Sea, in Koster, E. H., and Steel, R.I., eds., Sedimentology of gravels and conglomerates, Can. Soc. Petroleum Geol. Mem. 10, p. 383-398.

Kittredge, J., 1948, Forest influences: New York, McGraw-Hill, 394 p.

Kleinspehn, K. L., et al, 1984, Conglomeratic fan-delta sequences, Late Carboniferous-Early Permian, western Spitsbergen, in Koster, E. H., and Steel, R. J., eds., Sedimentology of gravels and conglomerates, Can. Soc. Petroleum Geol. Mem. 10, p. 279-294.

Knox, J. C., 1972, Valley alluviation in southwestern Wisconsin: Assoc. Am. Geography Ann., v. 62, p. 401-10.

environments of deposition and petroleum geology: Am. Assoc. Petroleum Geol. Bull., v. 69, p. 8981-898.

Mathisen, M. E., 1984, Diagenesis of Plio-Pleistocene nonmarine sandstones, Cagayon Basin, Phillipines: early development of secondary porosity in volcanic sandstones, in McDonald, D. A., and Surdam, R. C., eds., Clastic diagenesis: Amer. Assoc. Petroleum Geol. Mem. 37, p. 177-193.

McGowen, J. H., and Groat, C. G., 1971, Van Horn Sandstone, west Texas, an alluvial fan model for mineral exploration: Texas Bur. of Economic Geol. Rept. Inv. No. 72, 57 p.

McKee, E. D., et al, 1967, Flood deposits, Bijou Creek, Colorado: Jour. Sed. Petrology, v. 37, p. 829-851.

Melvin, John, and Knight, A. S., 1984, Lithofacies, diagenesis and porosity of the Ivishak Formation, Prudhoe Bay area, Alaska, in McDonald, D. A., and Surdam, R. C., eds., Clastic diagenesis: Amer. Assoc. of Petroleum Geol. Mem. 37, p. 347-366.

Miall, A. D., 1978, Tectonic setting and syndepositional setting of molasse and other non-marine-paralic sedimentary basins: Can. Jour. Earth Sci., v. 15, p. 1613-1632.

Miall, A. D., and M. R. Gibling, 1978, the Siluro-Devonian clastic wedge of Somerset Salard, Arctic Canada, and some regional paleogeographic implications: Sedimentary Geol., v. 21, p. 85-127.

Mitchell, A. H. G., and Reading, H. G., 1978, Sedimentation and tectonics, in Reading, H. G., ed., Sedimentary environments and facies: Oxford, Blackwell Scientific Publications, p. 439-476.

Nemac, W., and Steel, R. J., 1984, Alluvial and coastal conglomerates: their significant features and some comments on gravelly mass-flow deposits, in Koster, E. H., and Steel, R. J., eds., Sedimentology of gravels and conglomerates: Can. Soc. Petroleum Geol. Mem. 10, p. 1-32.

Nilsen, T. H., 1969, Old Red sedimentation in the Buelandet-Vaerlandet Devonian district, western Norway: Sedimentary Geology, vol. 3, p. 35-57.

Nilsen, T. H., and Moore, T. E., 1984, Bibliography of alluvial-fan deposits: Norwich, Geo Books, 96 p.

Koster, E. H., and Steel, R. J., eds., Sedimentology gravels and conglomerates: Can. Soc. Petroleum Geol. Mem. 10, 441 p.

Langbein, W. B., et al, 1949, Annual runoff in the United States: U. S. Geol. Survey Circ. 52, 14 p.

Langbein, W. B., and Schumm, S. A., 1958, Yield of sediment in relation to mean annual precipitation: Am. Geophys. Union Trans., v. 39, p. 1076-1084.

Larsen, V., and Steel, R. J., 1978, The sedimentary history of a debris-flow dominated alluvial fan--a study of textural inversion: Sedimentology, v. 25, p. 37-59.

Lattman, L. H., 1973, Calcium carbonate cementation of alluvial fans in southern Nevada: Geol. Soc. America Bull., v. 84, p. 3013-3028.

Link, M. H., 1982, Introduction to the facies of the Ridge Route Formation, Ridge Basin, southern California: in Crowell, J. C., and M. H. Link, eds., Geologic history of the Ridge Basin, southern California, Pacific Sec. S.E.P.M., Los Angeles, p. 99-104.

Link, M. H., and R. L. Osborne, 1982, Sedimentary facies of Ridge Basin, southern California: in Crowell, J. C., and M. H. Link, eds., Geologic history of the Ridge Basin, southern California, Pacific Sec., S.E.P.M., Los Angeles, p. 63-78.

Lowe, D. R., 1979, Sediment gravity flows: their classification and some problems of application to natural flows and deposits, in Doyle, L. J., and Pilkey, O. H., eds., Geology of continental slopes, S. E. P. M. Spec. Pub. 27, p. 75-82.

Lowe, D. R., 1982, Sediment gravity flows: II. Depositional models with special reference to the deposits of high-density turbidity currents: Jour. Sed. Petrology, v. 52, p. 279-297.

Mack, G. H., and Rasmussen, K. A., 1984, Alluvial-fan sedimentation of the Cutler Formation (Permo-Pennsylvanian) near Gateway, Colorado: Geol. Soc. America Bull., v. 95, p. 109-116.

Macke, D., 1976, Stratigraphy and sedimentology of experimental alluvial fans: Unpub. M. S. Thesis, Colorado State Univ., 105 p.

Mancini, E. A., et al, 1985, Norphlet Formation (Upper Jurassic) of southwestern and offshore Alabama:

Parker, R. S., 1976, Experimental study of drainage system evolution: unpub. report, Colorado State Univ.

Parker, R. S., 1977, Experimental study of drainage basin evolution and its hydrologic implications: Colorado State Univ. Hydrology Paper 90.

Patton, D. C., and Schumm, S. A., 1975, Gully erosion, northern Colorado: a threshold phenomenon: Geology, v. 3, p. 88-90.

Picard, M. D., and High, L. R., 1973, Sedimentary structures of ephemeral streams: Developments in Sedimentology No. 17, Amsterdam, Elsevier, 223 p.

Postma, G., 1984, Mass-flow conglomerates in a submarine canyon: Abrioja fan-delta, Pliocene, southern Spain, in Koster, E. H., and Steel, R. J., eds. Sedimentology of gravels and conglomerates. Can. Soc. Petroleum Geol. Mem. 10, p. 237-258.

Postma, George, and Roep, T. B., 1985, Resedimented conglomerates in the bottomsets of Gilbert-type gravel deltas: Jour. Sed. Petrology, v. 55, p. 874-885.

Pretorius, D. A., 1979, The depositional environment of the Witwatersrand goldfields: a chronological review of speculations and observations, in Anderson, A. M., and Biljon, W. J., eds., Some sedimentary basins and associated ore deposits of South Africa: Geol. Soc. South Africa Spec. Pub. No. 6, p. 33-55.

Reading, H. G., 1980, Characteristics and recognition of strike-slip fault systems; in Ballance, P. F., and Reading, H. G., eds., Sedimentation in Oblique-slip Mobile Zones: Int. Assoc. Sedimentologist, Spec. Pub. 4, p. 7-26.

Reid, J. C., 1974, Hazel Formation, Culberson and Hudspeth Counties, Texas: Unpub. M. S. Thesis, Univ. of Texas, Austin, 88 p.

Retallack, G. J., 1976, Triassic paleosols in the Upper Narabeen Group of New Southwales, Part I: Feature of the paleosols: Jour. Geol. Soc. Australia, v. 23, p. 383-399.

Retallack, G. J., 1985, Fossil soils as grounds for interpreting the advent of large plants and animals on land: Phil. Trans. Royal Soc. London, v. 309, p. 105-142.

Riba, O., 1976, Syntectonic unconformities of the Alto Cardener, Spanish Pyrenees: a genetic interpretaion: Sedimentary Geol., v. 15, p. 213-233.

Roe, S.-L., and Steel, Ron, 1985, Sedimentation, sea-level rise and tectonics at the Triassic-Jurassic boundary (Stratfjord Formation), Tampen Spur, northern North Sea: Jour. Petroleum Geol., v. 8, p. 163-186.

Russel, R. J., 1956, Environmental changes through forces independent of man: in Thomas, W. L., Jr., ed., Man's role in changing the face of the earth: Chicago, Univ. of Chicago Press, p. 453-470.

Rust, B. R., 1978, Depositional models for braided alluvium, in Miall, A. D., ed., Fluvial sedimentology: Can. Soc. Petroleum Geol. Mem. 5, p. 605-625.

Schumm, S. A., 1963, A tentative classification of alluvial river channels: U. S. Geol. Survey Circ. 477, 10 p.

Schumm, S. A., 1965, Quaternary paleohydrology, in Wright, H. E., and Frey, D. G., eds., Quaternary of the United States: Princeton, Princeton Univ. Press, p. 783-794.

Schumm, S. A., 1968, Speculations concerning paleohydrologic controls of terrestrial sedimentation: Geol. Soc. America Bull., v. 79, p. 1573-1588.

Schumm, S. A., 1977, The fluvial system: New York, John Wiley and Sons, 338 p.

Seni, S. J., 1980, Sand-body geometry and depositional systems, Ogallala Formation, Texas: Texas Bur. of Economic Geol. Rept. Inv., No. 105, 36 p.

Sestini, G., 1970, Postgeosynclinal deposition: Sedimentary Geol., v. 4, p. 481-520.

Sharp, R. P., 1948, Early Tertiary fan conglomerate, Big Horn Mountains, Wyoming: Jour. Geology v. 56, p. 1-15.

Shelby, J. M., 1980, Geologic and economic significance of the Upper Morrow Chert Conglomerate Reservoir of the Anardarko Basin: Jour. Petroleum Tech., v. 32, p. 489-495.

Shultz, A. W., 1983, Discussion: The deposits, internal structure, and geometry in six alluvial fan-fan delta bodies (Devonian-Norway) - a study in the significance of bedding sequences in conglomerates: Jour. Sed. Petrology, v. 53, p. 325-327.

Shultz, A. W., 1984, Subaerial debris-flow deposition in the Upper Paleozoic Cutler Formation, western Colorado: Jour. Sed. Petrology, v. 54, p. 759-772.

Smith, G. A., 1986, Coarse-grained nonmarine volcaniclastic sediment: terminology and depositional process: Geol. Soc. America Bull., v. 97, p. 1-10.

Stanton, G. O., 1977, Secondary porosity in sandstones of the lower Wilcox (Eocene), Karnes County, Texas: Gulf Coast Assoc. Geol. Soc. Trans., v. 27, p. 197-207.

Steel, R. J., 1974, New Red Sandstone floodplain and piedmont sedimentation in the Hebridean province, Scotland: Jour. Sed. Petrology, v. 44, p. 336-57.

Steel, R. J., and A. C. Wilson, 1975, Sedimentation and tectonism (?Permo-Triassic) on the margin of the North Minch Basin, Lewis: Jour. Geol. Soc. London, v. 131, p. 183-202.

Steel, R. J., 1976, Devonian basins of western Norway--sedimentary responses to tectonism and to varying tectonic context: Tectonophysics, v. 36, p. 207-224.

Steel, R. J., et al, 1977, Coarsening-upward cycles in the alluvium of Hornelen Basin (Devonian) Norway: sedimentary response to tectonic events: Geol. Soc. America Bull, v. 88, p. 1124-1134.

Stonecipher, S. A., Winn, R. D., Jr., and Bishop, M. G., 1984, Diagenesis of the Frontier Formation, Moxa Arch.: a function of sandstone geometry, texture and composition, and fluid flux, in McDonald, D. A., and Surdam, R. C., eds., Clastic Diagenesis: Amer. Assoc. of Petroleum Geol. Mem. 37, p. 289-317.

Strand, R. I., 1975, Bureau of Reclamation procedures for predicting sediment yield: Agricultural Res. Serv. ARS-S-40, p. 10-15.

Surdam, R. C., and Boles, J. R., 1979, Diagenesis of volcanic sandstones, in Scholle, P. A., and Schluger, P. R., eds., Aspects of Diagenesis: S. E. P. M. Spec. Pub. 26, p. 227-242.

Suttner, L. J., Basu, A., and Mack, C. H., 1981, Climate and the origin of quartz arenite: Jour. Sed. Petrology, v. 51, p. 1235-1246.

Suttner, L. J., Langford, R. D., and Mack, G. H., 1985, Guide to the field study of alluvial fan and fan-delta deposits in the Fountain Formation (Pennsylvanian-Permian), Colorado, in Flores, R. M., and Harvey, Michael, eds., Field guidebook to modern and ancient fluvial systems in the United States: Third Int. Fluvial `Sed. Conf. Proc., p. 65-84.

Suttner, L. J., and Dutta, P. K., 1986, Alluvial sandstone composition and paleoclimate: Part I, Framework mineralogy: Jour. Sed. Petrology, v. 56, (in press).

Taylor, J. C. M., 1978, Control of diagenesis by depositional environment within a fluvial sandstone sequence in the northern North Sea Basin: Jour. Geol. Soc. London, v. 135, p. 83-91.

Tilley, B. J., and Longstaffe, F. V., 1984, Controls on hydrocarbon accumulation in glauconitic sandstone, Suffield heavy oil sands, Southern Alberta: Amer. Assoc. Petroleum Geol. Bull., v. 68, p. 1004-1023.

Varley, C. J., 1984, Sedimentology and hydrocarbon distribution of the Lower Cretaceous Cadomin Formation, northwest Alberta, in Koster, E. H., and Steel, R. J., eds., Sedimentology of gravels and conglomerates, Can. Soc. Petroleum Geol. Mem. 10, p. 175-188.

Walker, T. R., Waugh, Brian, and Grone, A. J., 1978, Diagenesis in first-cycle desert alluvium of Cenozoic age, southwestern United States and northwestern Mexico: Geol. Soc. America Bull., v. 89, p. 19-32.

Weaver, W. E., 1976, Experimental study of alluvial fans: Unpub. Ph. D. Dissertation, Colorado State Univ.

Wells, S. G., and Dohrenwend, J. C., 1985, Relict sheetflood bed forms on Late Quaternary alluvial-fan surfaces in the southwestern United States: Geology, v. 13, p. 512-516.

Wescott, W. A., and Ethridge, F. G., 1984, Fan-delta sedimentology and tectonic setting--Yallahs Fan Delta, southeast Jamaica: Am. Assoc. Petroleum Geol. Bull., v. 64, p. 374-399.

Wilson, A. C., 1980, The Devonian sedimentation and tectonism of a rapidly subsiding, semi-arid fluvial basin in the Midland Valley of Scotland: Scott. Jour. Geol., v. 16, p. 291-313.

Wise, D. U., 1963, Keystone faulting and gravity sliding driven by basement uplift of Owl Creek Mountains, Wyoming: Am. Assoc. Petroleum Geol., v. 47, p. 586-598.

ADDITIONAL REFERENCES

The references listed below are not cited in the text of this book but are, nonetheless, important recent contributions that we consider essential to an understanding of the sedimentology of alluvial fans and fan deltas.

Allen, P. A., 1981, Sediments and processes on a small stream-flow dominated, Devonian alluvial fan, Shetland Islands: Sedimentary Geol., v. 29, p. 31-66.

Allen, P. A., and Marshall, J. E. A., 1981, Depositional environments and palynology of the Devonian south-east Shetland Basin: Scottish Jour. Geol., v. 17, p. 257-273.

Allen, P. A., *et al*, 1983, Variations in fluvial style on the Eocene-Oligocene alluvial fan of the Scala Dei Group, SE Ebro Basin, Spain: Jour. Geol. Soc. London, v. 140, p. 133-146.

Arlucea, M. P., and Sopena, A., 1983, Estudio estratigraphico y sedimentologico de los materiales Permicos y Triasicos en el noroeste de la Sierra de Albarracin (provincia de Guadalajara): Estudios Geol., v. 39, p. 329-343.

Blake, T. F., 1982, Depositional environments of the Simmler Formation in southern Cuyama Valley, Santa Barbara and Ventura Counties, California, *in* Ingersoll, R. V., and Woodburne, M. O., eds., Cenozoic nonmarine deposits of California and Arizona: Pacific Section, S. E. P. M., p. 35-50.

Bohannon, R. G., 1984, Nonmarine sedimentary rocks of Tertiary age in the Lake Mead region, southeastern Nevada and northwestern Arizona: U. S. Geol. Survey, Prof. paper 1259, 72 p.

Brookfield, M. E., and Andrews-Speed, C. P., 1984, Sedimentology, petrography and tectonic significance of the shelf, flysch and molasse clastic deposits across the Indus suture zone, Ladakh, NW India: Sedimentary Geol., v. 40, p. 249-286.

Buck, S. G., 1980, Stromatolite and ooid deposits within the fluvial and lacustrine sediments of the Precambrian Ventersdorp Supergroup of South Africa: Precambrian Res., v. 12, p. 311-330.

Burgisser, H. M., 1981, Facies and Palaohydrologie der Oberen Susswassermolasse im Hornli-Facher (Nordostschweiz): Eclogae Geol. Helv., v. 74, p. 19-28.

Costa, J. E., and Jarrett, R. D., 1981, Debris flows in small mountain stream channels of Colorado and their hydrologic implications: Assoc. Engineering Geol. Bull., v. 18, p. 309-322.

Farquharson, G. W., 1984, Late Mesozoic, non-marine conglomeratic sequences of northern Antarctic Peninsula (the Botany Bay Group): Brit. Antarctic Surv. Bull. No. 65, p. 1-32.

Feehan, J., 1980, Alluvial fan sediments from the Old Red Sandstone of Devilsbit Mountain, County Tipperary: Jour. Earth Sci. Royal Dublin Soc., v. 3, p. 179-194.

Frostick, L. E., and Reid, I., 1980, Sorting mechanisms in coarse-grained alluvial sediments: fresh evidence from a basalt plateau gravel, Kenya: Jour. Geol. Soc. London, v. 137, p. 431-441.

Graham, J. R., 1981, The 'Old Red Sandstone' of County Mayo, northwest Ireland: Geological Jour., v. 16, p. 157-173.

Halaf, F. I., Gharib, I. M., and Al Kadi, A. S., 1982, Sources and genesis of the Pleistocene gravelly deposits in northern Kuwait: Sedimentary Geol., v. 31, p. 101-117.

Hayward, A. B., 1985, Coastal alluvial fans (fan deltas) of the gulf of Aqaba (Gulf of Eilat), Red Sea: Sedimentary Geol., v. 43, p. 241-260.

Hiller, Norton, and Stavrakis, Nicholas, 1984, Permo-Triassic fluvial systems in the southeastern Karoo Basin, South Africa: Palaeogeography Palaeoclimatal. Palaeoecol., v. 45, p. 1-21.

Hogg, S. E., 1982, Sheetfloods, sheetwash, sheetflow, or.....?: Earth-Sci. Rev., v. 18, p. 59-76.

Houmark-Nielsen, Michael, 1983, Depositional features of a late Weichselian outwash fan; central East Jylland, Denmark: Sedimentary Geol., v. 36, p. 51-63.

Mader, Detlef, 1981, Genesis of the Buntsandstein (Lower Triassic) in the Western Eifel (Germany): Sedimentary Geol., v. 29, p. 1-30.

Mader, Detlef, 1983, Evolution of fluvial sedimentation in the Buntsandstein (Lower Triassic) of the Eifel (Germany): Sedimentary Geol., v. 37, p. 1-84.

Madole, R. F., 1982, Possible origins of till-like deposits near the summit of the Front Range in north-central Colorado: U. S. Geol. Survey Prof. Paper 1243, 31 p.

Mills, H. H., 1982, Long-term episodic deposition on mountain foot slopes in the Blue Ridge Province of North Carolina: evidence from relative-age dating: Southeastern Geol., v. 23, p. 123-128.

Minter, W. E. L., 1981, Examples that illustrate sedimentological aspects of the Proterozoic placer model on the Kaap-Vaal Craton, Witwatersrand, South Africa Genesis of uranium-and gold-bearing Precambrian quartz-pebble conglomerates : U. S. Geol. Survey Prof. Paper 1161 E, 6 p.

Nio, S.-D., and Hussain, S. T., 1984, Sedimentological framework of Late Pliocene and Pleistocene alluvial deposits in the Bhittani Range, Pakistan: Geol. Mijnbouw, v. 63, p. 55-70.

Nystuen, J. P., 1982, Late Proterozoic basin evolution of the Baltoscandian Craton: the Hedmark Group, southern Norway: Norges Geol. Unders. No. 375, p. 1-74.

Ori, G. G., and Penney, S. R., 1982, The Stratigraphy and sedimentology of the Old Red Sandstone sequence at Dunmore East, County Waterford: Jour. Earth Sci. Royal Dublin Soc., v. 5, p. 43-59.

Palmer, C. M., and Merrill, R. D., 1982, Braided-stream and alluvial-fan depositional environments in the lower to middle Eocene Ione Formation, Madera county, California, in Ingersoll, R. V., and Woodburne, M. O., eds., Cenozoic nonmarine deposits of California and Arizona: Pacific Section S.E.P.M., p. 1-10.

Parkash, B., Sharma, R. P., and Roy, A. K., 1980, The Siwalik Group (molasse) - sediments shed by collision of continental plates: Sedimentary Geol., v. 25, p. 127-159.

Phillips, W. E. A., and Clayton, G., 1980, The Dinantian clastic succession of Clare Island, County Mayo: Jour. Earth Sci. Royal Dublin Soc., v. 2, p. 115-135.

Pomeroy, J. S., 1983, Relict debris flows in northwestern Pennsylvania: Northeastern Geol., v. 5, p. 1-7.

Ramos, Amparo, and Friend, P. F., 1982, Upper Old Red Sandstone sedimentation near the unconformity at Arbroath: Scottish Jour. Geol., v. 18, p. 297-315.

Roscoe, S. M., 1981, Temporal and other factors affecting deposition of uraniferous conglomerates in Genesis of uranium-and gold-bearing Precambrian quartz-pebble conglomerates : U. S. Geol. Survey Prof. Paper 1161 W, 17 p.

Rust, B. R., and Gostin, V. A., 1981, Fossil transverse ribs in Holocene alluvial fan deposits, Depot Creek, South Australia: Jour. Sed. Petrology, v. 51, p. 441-444.

Saller, A. H., and Dickinson, W. R., 1982, Alluvial to marine facies transition in the Antler overlap sequence, Pennsylvanian and Permian of north-central Nevada: Jour. Sed. Petrology, v. 52, p. 925-940.

Schneidegger, A. E., 1984, A review of recent work on mass movements on slopes and on rock falls: Earth-Sci. Rev., v. 21, p. 225-249.

Simon, J. B., 1984, Sedimentation of a small complex alluvial fan of possible Upper Old Red Sandstone age, northeast County Antrim: Irish Jour. Earth Sci., v. 6, p. 109-119.

Skinner, B. J., 1981, Thoughts about uranium-bearing quartz-pebble conglomerates: a summary of ideas presented at the workshop in genesis of uranium-and gold-bearing Precambrian quartz-pebble conglomerates : U. S. Geol. Survey Prof. Paper 1161 BB, 5 p.

Squires, R. L., and Advocate, D. M., 1982, Sedimentary facies of the nonmarine Lower Miocene Diligencia Formation, Canyon Spring area, Orocopia Mountains, southern California, in Ingersoll, R. V., and Woodburne, M. O., eds., Cenozoic nonmarine deposits of California and Arizona: Pacific Section S. E. P. M., p. 101-106.

Stoker, M. S., 1982, Old Red Sandstone sedimentation and deformation in the Great Glen Fault Zone, NW of Loch Linnhe: Scottish Jour. Geol., v. 18, p. 147-156.

Suguio, Kenitiro, and Barcelos, J. H., 1983, Paleoclimatic evidence from the Areado Formation, Cretaceous of the Sao Francisco Basin, state of Minas Gesais, Brazil: Rev. Brasileira Geoc., v. 13, p. 229-231.

Suguio, Kenitiro, and Barcelos, J. H., 1983, Paleoclimatic evidence from the Bauru Group, Cretaceous of the Parana Basin, Brazil: Rev. Brasileira Geoc., v. 13, p. 232-236.

Theakstone, W. H., 1982, Sediment fans and sediment flows generated by snowmelt: observations at Austerdalsisen, Norway: Jour. Geology, v. 90, p. 583-588.

Tunbridge, I. P., 1983, Alluvial fan sedimentation of the Horseshoe Park flood, Colorado, U. S. A., July 15th, 1982: Sedimentary Geol., v. 36, p. 15-23.

Index